Georg Keller

Realistic Modeling of Strongly Correlated Electron Systems

λογος

Augsburger Schriften zur Mathematik, Physik und Informatik

Band 6

herausgegeben von:
Professor Dr. F. Pukelsheim
Professor Dr. W. Reif
Professor Dr. D. Vollhardt

Bibliografische Information Der Deutschen Bibliothek

Die Deutsche Bibliothek verzeichnet diese Publikation in der Deutschen Nationalbibliografie; detaillierte bibliografische Daten sind im Internet über http://dnb.ddb.de abrufbar.

ISBN 3-8325-0970-4

ISSN 1611-4256

Logos Verlag Berlin
Comeniushof, Gubener Str. 47,
10243 Berlin
Tel.: +49 030 42 85 10 90
Fax: +49 030 42 85 10 92
INTERNET: http://www.logos-verlag.de

Realistic Modeling of Strongly Correlated Electron Systems

Von der Mathematisch-Naturwissenschaftlichen Fakultät
der Universität Augsburg
zur Erlangung eines Doktorgrades der Naturwissenschaften
genehmigte Dissertation

von
Diplom–Physiker Georg Keller
aus
Füssen

Erstgutachter:	Prof. Dr. D. Vollhardt
Zweitgutachter:	Prof. Dr. Th. Kopp, Prof. Dr. Th. Pruschke

Tag der mündlichen Prüfung: 10. Juni 2005

CONTENTS

INTRODUCTION

The physics of materials with strongly correlated electrons is one of the most exciting topics of present-day theoretical and experimental solid-state research. A wide variety of interesting phenomena can be attributed to electronic correlations, among them metal-insulator transitions, the giant and colossal magnetoresistance effect, superconductivity and heavy-fermion behavior. One focus of experimental research is on transition metals and especially transition metal oxides due to the abundance of materials and the diversity of phenomena encountered in these systems. The electronic structure of such materials is strongly influenced by the partially filled d-orbitals, which have only a small direct overlap but still have itinerant character due to the indirect overlap via oxygen p-orbitals. In transition metal oxides, the efficiency of the screening and thus the strength of the correlations depends crucially on the position of the (empty) 4s-band and the (partially filled) 3d-bands. Furthermore, the crystal field splitting of the five d-orbitals due to the influence of the neighboring (oxygen) atoms, which is often of the same magnitude as the bandwidth, plays an important role for the physical properties of this class of systems.

An accurate theoretical description of such strongly correlated systems is exceedingly difficult. Density functional calculations (e.g., in the local density approximation LDA) often only provide a rough characterization of those materials and even give qualitatively wrong results in many cases, e.g., they predict that Cr-doped V_2O_3 is metallic whereas it is an insulator in experiments. Clearly, band structure methods are not sufficient for strongly correlated materials and a genuine many-body approach is necessary. However, those many-body methods are feasible mainly for simplified effective models. Although such models can capture certain aspects of the underlying physics and allow for a qualitative description of real systems, they depend on model parameters that have to be adjusted manually and cannot provide a realistic *ab initio* description. One of the most successful model approaches in this context is the dynamical mean-field theory (DMFT). It is a non-perturbative method for solving models with strongly correlated electrons, especially the Hubbard model and its extensions. DMFT becomes exact in the limit of infinite dimensions or infinite lattice coordination number (Metzner and Vollhardt 1989).

A novel approach that was developed in the last few years and has already lead to impressive results is the LDA+DMFT method (Anisimov, Poteryaev, Korotin, Anokhin and Kotliar 1997). It combines the advantages of the LDA, which provides a realistic *ab initio* description for many materials, with the correct treatment of the local correlations within DMFT. A robust and numerically exact impurity solver for the effective quantum impurity problem encountered in DMFT is the quantum Monte Carlo (QMC) algorithm. It can be applied for multi-orbital systems like transition metal oxides with five 3d-orbitals or rare earth materials with seven 4f-orbitals for not too low temperatures. Thus, the LDA+DMFT method makes it possible for the first time to perform parameter-free *ab initio* calculations for strongly correlated materials (Vollhardt, Held, Keller, Bulla, Pruschke, Nekrasov and Anisimov 2005) and gives new insight into long-debated problems like the metal-insulator transition in V_2O_3 (Held, Keller, Eyert, Anisimov and Vollhardt 2001).

The main topic of this thesis is the investigation of strongly correlated transition metal oxide systems with the LDA+DMFT approach, where the self-consistent equations of the DMFT are solved with an auxiliary-field quantum Monte Carlo algorithm (Hirsch and Fye 1986). In particular, the metallic and insulating phase of V_2O_3 and the peculiarities of its metal-insulator transition are explored. Furthermore, the strongly correlated metals $SrVO_3$ and $CaVO_3$ are studied, as well as LiV_2O_4, the first d-electron system that was found to exhibit heavy-fermion behavior. Where possible, the theoretical results are compared with recent experimental data.

Outline of this thesis

In chapter 1, the *ab initio* electronic Hamiltonian is introduced. The different approximation strategies for the full correlation problem, i.e., the density functional approach and the model approach, are discussed. In the first part of the chapter, the density functional theory and its local density approximation (LDA) are presented and the inclusion of local Coulomb correlations in the framework of the LDA Hamiltonian is demonstrated. Furthermore, the constrained LDA is described which allows for a calculation of the Coulomb interaction parameters in the framework of the local density approximation. In the second part, one of the simplest models of many-body theory, the Hubbard model and its extension to multiple bands is presented. After that, the dynamical mean-field theory is reviewed and its application to the LDA Hamiltonian with local correlations is shown. The simplifications that arise for systems with degenerate bands are studied and the corresponding Dyson equation is given. Finally, some extensions to the LDA+DMFT scheme (the Wannier function formalism, a self-consistent LDA+DMFT algorithm and cluster extensions to DMFT) are presented that make more advanced calculations possible.

In chapter 2, the numerical aspects of solving the DMFT equations are discussed. The auxiliary-field quantum Monte Carlo (QMC) algorithm, which is the basis of the calculations presented in this work, is reviewed and the problem of the Fourier transformations within DMFT(QMC) is studied. Furthermore, the maximum entropy method, which allows to obtain spectral functions on the real axis from the imaginary-time data of QMC, is discussed. The chapter closes with a description of the computational aspects of the QMC, the parallelization of the code and a performance comparison of the parallel and sequential code on different computer architectures.

V_2O_3, the most famous example for a system that shows a correlation induced metal-insulator transition, is studied in chapter 3. After the crystal and electronic structure and the results from LDA calculations, the spectra obtained with LDA+DMFT are presented. The local magnetic moment and the orbital occupation across the Mott-Hubbard transition are explored and the changes of the spectral weight at the Fermi edge due to the quasiparticle renormalization are discussed. A detailed comparison of the theoretical spectra to experimental photoemission (PES) and x-ray absorption (XAS) measurements is made.

In chapter 4, the strongly correlated systems $SrVO_3$ and $CaVO_3$ are investigated. Starting with the crystal and electronic structure of those systems, the densities of states obtained by LDA are studied. The spectra from the subsequent LDA+DMFT calculations show that both materials are correlated metals and far from a metal-insulator transition. Good agreement is found in the comparison with the experimental spectra from high-resolution photoemission and x-ray absorption spectroscopy.

The 3d heavy-fermion system LiV_2O_4 is studied in chapter 5. After a discussion of the crystal and electronic structure of the material, the results from a LDA calculation are presented. Next, the correlated spectra obtained with LDA+DMFT are studied and the origin of the peaks is analyzed based on a NCA (non-crossing approximation) calculation. The magnetic properties of LiV_2O_4 and the different exchange contributions are explored and the theoretical data is compared with experiments.

Chapters 3, 4 and 5 each contain a conclusion. In the summary, a brief overview of the most important results achieved in this work is presented.

1. THE LDA+DMFT APPROACH

In many systems that are studied in solid-state physics, the electronic properties are well described by the electronic Hamiltonian

$$\hat{H} = \sum_{\sigma} \int d^3r\, \hat{\Psi}^+(\boldsymbol{r},\sigma) \left[-\frac{\hbar^2}{2m_e}\Delta + V_{\text{ion}}(\boldsymbol{r}) \right] \hat{\Psi}(\boldsymbol{r},\sigma)$$
$$+\frac{1}{2}\sum_{\sigma\sigma'} \int d^3r\, d^3r'\, \hat{\Psi}^+(\boldsymbol{r},\sigma)\hat{\Psi}^+(\boldsymbol{r}',\sigma')\, V_{\text{ee}}(\boldsymbol{r}-\boldsymbol{r}')\, \hat{\Psi}(\boldsymbol{r}',\sigma')\hat{\Psi}(\boldsymbol{r},\sigma). \quad (1.1)$$

Here, the crystal lattice enters only via

$$V_{\text{ion}}(\boldsymbol{r}) = -e^2 \sum_i \frac{Z_i}{|\boldsymbol{r}-\boldsymbol{R}_i|}, \quad (1.2)$$

which denotes the one-particle ionic potential of all ions i with charge eZ_i at given positions $\boldsymbol{R}_i$. $\hat{\Psi}^+(\boldsymbol{r},\sigma)$ and $\hat{\Psi}(\boldsymbol{r},\sigma)$ are field operators that create and annihilate an electron at position $\boldsymbol{r}$ with spin σ, Δ is the Laplace operator, m_e the electron mass, e the electron charge and

$$V_{\text{ee}}(\boldsymbol{r}-\boldsymbol{r}') = \frac{e^2}{2}\sum_{\boldsymbol{r}\neq\boldsymbol{r}'} \frac{1}{|\boldsymbol{r}-\boldsymbol{r}'|} \quad (1.3)$$

the electron-electron interaction. The Hamiltonian of Eq. (1.1) is valid within Born-Oppenheimer approximation (Born and Oppenheimer 1927) and does not include relativistic effects. An exact solution of the *ab initio* Hamiltonian is only possible for a very small number of electrons. Even numerical methods like the Green-function Monte Carlo and related approaches have been used successfully only for relatively modest numbers of electrons, i.e., for light atoms and molecules like hydrogen and helium. The reason for this is the electron-electron interaction (1.3) which correlates every electron with all others and leads to a exponential increase in the complexity of the problem with the number of electrons. In order to overcome this complexity, one can either make substantial approximations to deal with the Hamiltonian (1.1) or replace it with a strongly simplified model Hamiltonian. Both strategies are applied for the investigation of electronic properties in solids, in the density functional theory (DFT) and in many-body theory, respectively. The DFT, especially in the local density approximation (LDA), is a highly successful method for realistic solid-state calculations (Jones and Gunnarsson 1989). Although in LDA, the correlations and also the exchange contribution of the Coulomb interaction are only treated rudimentarily (see Sec. 1.1 for details), experience shows that both ground-state energies and band structures can be calculated accurately in LDA for many, weakly correlated systems. However, in materials with partly filled d- or f-electron shells where the Coulomb interaction is comparable to the bandwidth, i.e., in strongly correlated systems (like transition metal oxides or heavy-fermion systems), the LDA is seriously restricted in its accuracy and sometimes even produces qualitatively wrong results. For example, for V_2O_3 and La_2CuO_4 which are antiferromagnetic insulators at low temperatures, LDA predicts non-magnetic metallic behavior (Leung, Wang and Harmon 1988, Zaanen, Jepsen, Gunnarsson, Paxton, Andersen and Svane 1988, Pickett 1989, Mattheiss 1994). In such correlation-induced insulators (so-called Mott insulators), the physics is dominated by the splitting of the LDA bands into two bands which are separated by a Coulomb repulsion U. The formation of those so-called Hubbard bands is associated with an energy gain of the same order of magnitude. The Mott insulating behavior is physically independent of the magnetic ordering (e.g., V_2O_3 has a paramagnetic Mott insulating phase). The ordering is a secondary effect and normally sets in at lower temperatures, but it can also occur simultaneously with the metal-insulator transition (as in the transition from the paramagnetic metallic to antiferromagnetic insulating phase in V_2O_3). For ordered systems, the Mott physics that is completely missing in LDA can be described by the LDA+U method (Anisimov, Zaanen and Andersen 1991, Liechtenstein, Anisimov and Zaanen 1995). A major drawback of this method is that the energy of the correlated metal is strongly overestimated. Hence, for realistic values of the Coulomb interaction U, the energy gain due to the Hubbard band formation is so large that LDA+U almost automatically predicts insulating behavior, in some cases even for metallic systems. It is therefore well suited for the description of ordered Mott insulators but it fails for strongly correlated metals or for systems near a Mott-Hubbard metal-insulator transition.

Neither LDA nor LDA+U are able to describe the Kondo-like energy scale associated with the quasiparticle physics. The formation of quasiparticles with a larger effective mass than that of the LDA electrons[1] leads to a redistribution of spectral weight away from the Fermi edge, but the metallic behavior is still retained up to rather large Coulomb interaction U. In this respect, the model Hamiltonian approach is much more powerful than LDA and LDA+U. Various techniques exist to investigate the many-body problem. With those techniques,

[1] This mass enhancement can range from a relatively small increase ($m^* \approx 2 - 5m_e$) in many transition metal oxides to the huge effective masses ($m^* \approx 100 - 1000m_e$) found in heavy-fermion systems with 4f-electrons.

it is possible to describe the qualitative features of the models under investigation and understand the basic mechanisms of various physical phenomena, e.g., the formation of quasiparticles or of a Mott-Hubbard gap. One of the most successful many-body techniques for the non-perturbative modelling of strongly correlated electron systems is the dynamical mean-field theory (DMFT). In Sec. 1.2.2, the method and the various numerical techniques to solve it are presented. A major drawback of model Hamiltonian approaches is that input parameters are not a priory known and have to be adjusted, hence quantitative predictions for real materials are possible only to a limited extent.

The LDA+DMFT combines the positive aspects of DFT/LDA and DMFT. This technique was first formulated by Anisimov, Poteryaev, Korotin, Anokhin and Kotliar (1997) and since then developed further by various groups.[2] In the LDA+DMFT scheme, the weakly correlated part of the *ab initio* Hamiltonian (1.1), i.e., the electrons in the s- and p-orbitals and the long-range interaction of the electrons in d- and f-orbitals, is described by the LDA. The DMFT is used to describe the strong correlations between the electrons in the d- and f-shells due to the local Coulomb interaction. Thus, LDA+DMFT provides a realistic description for weakly as well as strongly correlated materials. For small Coulomb interaction, the LDA as well as the LDA+DMFT describe a weakly correlated metal as denoted by the density of states in the left part of Fig. 1.1.[3] At large Coulomb interaction, the LDA+DMFT is in agreement with LDA+U results for Mott insulators with symmetry breaking. Both methods show a density of states with a lower and upper Hubbard band and a Mott gap $\sim U$ (right part of Fig. 1.1). Furthermore, LDA+DMFT can describe the formation of a quasiparticle peak at intermediate interaction strengths and thus provide a correct description of correlated metals and doped Mott insulators (see middle of Fig. 1.1).

A further advantage of the LDA+DMFT is that various techniques exist to solve the DMFT equations, among others quantum Monte Carlo (QMC) simulations, non-crossing approximation (NCA), iterative perturbation theory (IPT) and numerical renormalization group (NRG). In this work, only QMC will be described and used in the calculations.[4]

[2] For an overview of the method refer to (Held, Nekrasov, Blümer, Anisimov and Vollhardt 2001, Held, Nekrasov, Keller, Eyert, Blümer, McMahan, Scalettar, Pruschke, Anisimov and Vollhardt 2002, Held, Nekrasov, Keller, Eyert, Blümer, McMahan, Scalettar, Pruschke, Anisimov and Vollhardt 2003).

[3] In the case of LDA+DMFT solved with quantum Monte Carlo simulations, the finer structures of LDA are smoothed out due to the high temperature ($T \approx 300 - 1160$ K) used in the QMC calculations even for small values of U. Besides this QMC-specific effect, smoothing also occurs due to the correlations, i.e. due to the non-zero imaginary part of the self-energy $\text{Im}\Sigma(\omega)$.

[4] The use of NCA to solve the LDA+DMFT is shown in Held et al. (2002), further information on NRG and IPT can be found in, e.g., Bulla, Hewson and Pruschke (1998) and Georges, Kotliar, Krauth and Rozenberg (1996), respectively.

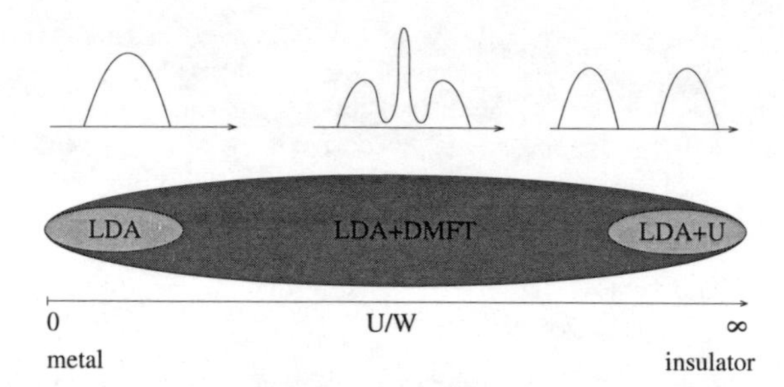

Figure 1.1: Evolution of the density of states with increasing U/W ratio (increasing Coulomb interaction U or decreasing bandwidth W (from Held et al. (2003)). LDA+DMFT provides a correct description for all values of U/W and includes the LDA results at weak interaction and the LDA+U results for large U/W.

Due to its ability to correctly and realistically describe materials ranging from weakly and strongly correlated metals to Mott insulators and the wide range of techniques that can be used for solving the DMFT equations, the LDA+DMFT has gained considerable interest in the many-body as well as the band structure community. Materials with strong correlations that could not be realistically described in LDA or with many-body techniques can now be treated with LDA+DMFT.

1.1 Density functional theory and local density approximation

1.1.1 Density functional theory

Density functional theory (DFT) has proven to be one of the most successful methods for realistic calculations in atoms, molecules and solids. Due to its wide range of applications, there exist many reviews on this subject, among others the reviews by Jones and Gunnarsson (1989) and by Capelle (2003). The DFT is based on the fact that the ground state of a system can be expressed as a functional of the electron density and that this functional assumes its minimum for the ground-state electron density. This theorem was first derived and proven by Hohenberg and Kohn (1964). A simpler and more general derivation was published by Levy (1979). It is valid for all electron densities $\rho(\boldsymbol{r})$ which can be obtained from an antisymmetric (many-body) wave function $\varphi(\boldsymbol{r}) = \varphi(\boldsymbol{r}_1\sigma_1, \boldsymbol{r}_2\sigma_2, \ldots, \boldsymbol{r}_N\sigma_N)$ with N electrons.[5] Taking the minimum (infimum) of the energy expectation value with respect to all wave functions $\varphi(\boldsymbol{r}_1\sigma_1, \boldsymbol{r}_2\sigma_2, \ldots, \boldsymbol{r}_N\sigma_N)$ which yield the density $\rho(\boldsymbol{r})$, the energy functional $E[\rho]$ can be obtained:

$$E[\rho] = \inf\left\{\langle\varphi|\hat{H}|\varphi\rangle \;\middle|\; \langle\varphi|\sum_{i=1}^{N}\delta(\boldsymbol{r}-\boldsymbol{r}_i)|\varphi\rangle = \rho(\boldsymbol{r})\right\}. \tag{1.4}$$

Since an evaluation of this functional would require an evaluation of the full Hamiltonian (1.1), this construction is of no direct practical value. The energy $E[\rho]$ can be expressed as a sum of contributions

$$E[\rho] = E_{\text{kin}}[\rho] + E_{\text{ion}}[\rho] + E_{\text{Hartree}}[\rho] + E_{\text{xc}}[\rho]. \tag{1.5}$$

Here, $E_{\text{kin}}[\rho]$ is the kinetic energy and $E_{\text{ion}}[\rho]$ the one-particle ionic potential (see the Hamiltonian (1.1) for comparison). The electron-electron interaction is split into the Hartree energy $E_{\text{Hartree}}[\rho]$ and the (unknown) exchange and correlation term $E_{\text{xc}}[\rho]$. Since the energy of the ionic potential $E_{\text{ion}}[\rho] = \int d^3r\, V_{\text{ion}}(\boldsymbol{r})\, \rho(\boldsymbol{r})$ and the Hartree energy $E_{\text{Hartree}}[\rho] = \frac{1}{2}\int d^3r'\, d^3r\, V_{\text{ee}}(\boldsymbol{r}-\boldsymbol{r}')\, \rho(\boldsymbol{r}')\rho(r)$ are known and can be directly written in terms of the electron density, all the difficulties of the many-body problem have been transferred into $E_{\text{xc}}[\rho]$. The kinetic energy $E_{\text{kin}}[\rho]$ cannot be explicitly expressed in terms of the electron density ρ, but one can employ the following procedure to determine it. Instead of minimizing the energy functional $E[\rho]$ with respect to ρ, it is minimized with respect to a set of one-particle wave functions φ_i which fulfill

$$\rho(\boldsymbol{r}) = \sum_{i=1}^{N}|\varphi_i(\boldsymbol{r})|^2. \tag{1.6}$$

The normalization of the $\varphi_i(\boldsymbol{r})$ is guaranteed by introducing the Lagrange parameters ε_i in the variation

$$\delta\{E[\rho] + \varepsilon_i[1 - \int d^3r|\varphi_i(\boldsymbol{r})|^2]\}/\delta\varphi_i(\boldsymbol{r}) = 0. \tag{1.7}$$

Finally, one obtains the Kohn-Sham equations (Kohn and Sham 1965, Sham and Kohn 1966)

$$\left[-\frac{\hbar^2}{2m_e}\Delta + V_{\text{ion}}(\boldsymbol{r}) + \int d^3r'\, V_{\text{ee}}(\boldsymbol{r}-\boldsymbol{r}')\rho(\boldsymbol{r}') + V_{\text{xc}}(\boldsymbol{r})\right]\varphi_i(\boldsymbol{r}) = \varepsilon_i\, \varphi_i(\boldsymbol{r}), \tag{1.8}$$

[5] The derivation by Hohenberg and Kohn (1964) was obtained in the space of densities that can be realized for an external potential $V_{\text{ext}}(\boldsymbol{r})$. This is only a subspace of the more general space of densities used by Levy (1979).

where the exchange correlation potential $V_{\rm xc}$ is the functional derivative of $E_{\rm xc}$:

$$V_{\rm xc}(\boldsymbol{r}) = \frac{\delta E_{\rm xc}[\rho]}{\delta\rho(\boldsymbol{r})}. \tag{1.9}$$

Here, the complicated many-body electron problem is converted to a set of non-interacting one-particle problems with a complicated (unknown) potential $V_{\rm xc}$. The ansatz with one-particle wave functions (1.6) can *a posteriori* be justified since the form of the Kohn-Sham equations (1.8) is identical to a one-particle Schrödinger equation. Now, one can obtain the kinetic energy for the ground-state density $\rho_{\rm min}$ as a sum of the kinetic energies of the one-particle wave functions

$$E_{\rm kin}[\rho_{\rm min}] = -\sum_{i=1}^{N}\langle\varphi_i|\hbar^2\Delta/(2m_e)|\varphi_i\rangle. \tag{1.10}$$

Here, the φ_i are the self-consistent (spin-degenerate) solution of (1.6) and (1.8) with lowest "energy" ε_i. It is important to note that the one-particle potential of the Kohn-Sham equations (1.8)

$$V_{\rm eff}(\boldsymbol{r}) = V_{\rm ion}(\boldsymbol{r}) + \int d^3r' V_{\rm ee}(\boldsymbol{r}-\boldsymbol{r}')\rho(\boldsymbol{r}') + V_{\rm xc}(\boldsymbol{r}), \tag{1.11}$$

is only an auxiliary potential that arises in the one-particle ansatz for the calculation of the kinetic energy. Thus, the wave functions φ_i and the Lagrange parameters ε_i have no physical meaning at this point.

Introducing an interaction parameter λ that is varied from 0 (non-interacting system) to 1 (physical system), the exchange correlation energy can be rewritten as

$$E_{\rm xc}[\rho] = \frac{1}{2}\int d^3r\rho(\boldsymbol{r})\int d^3r'\frac{1}{|\boldsymbol{r}-\boldsymbol{r}'|}\rho_{\rm xc}(\boldsymbol{r},\boldsymbol{r}'-\boldsymbol{r}), \tag{1.12}$$

with the exchange correlation hole[6]

$$\rho_{\rm xc}(\boldsymbol{r},\boldsymbol{r}'-\boldsymbol{r}) \equiv \rho(\boldsymbol{r}')\int_0^1 d\lambda[g(\boldsymbol{r},\boldsymbol{r}',\lambda)-1]. \tag{1.13}$$

$g(\boldsymbol{r},\boldsymbol{r}',\lambda)$ is the pair correlation function of the system with density $\rho(\boldsymbol{r})$ and Coulomb interaction $\lambda V_{\rm ee}$. From this form of the exchange correlation energy, some important insight can be gained. Since $g(\boldsymbol{r},\boldsymbol{r}',\lambda) \to 1$ for $|\boldsymbol{r}-\boldsymbol{r}'| \to \infty$, the separation of $E_{\rm xc}[\rho]$ into an electrostatic ($\frac{1}{|\boldsymbol{r}-\boldsymbol{r}'|}$) and exchange correlation ($\rho_{\rm xc}(\boldsymbol{r},\boldsymbol{r}'-\boldsymbol{r})$) contribution in Eq. (1.12) can also be viewed as an approximate separation of the Coulomb energy into a long-range and short-range contribution,

[6] The exchange correlation hole $\rho_{\rm xc}(\boldsymbol{r})$ describes the fact that an electron at point $\boldsymbol{r}$ reduces the probability of finding an electron at point $\boldsymbol{r}'$. Thus, $E_{\rm xc}[\rho]$ can be viewed as the energy which results from the interaction of an electron with its exchange correlation hole.

First principles information: atomic numbers, crystal structure (lattice, atomic positions)
Choose initial electronic density $\rho(\boldsymbol{r})$
Calculate effective potential using the LDA [Eq. (1.11)] $$V_{\text{eff}}(\boldsymbol{r}) = V_{\text{ion}}(\boldsymbol{r}) + \int d^3\boldsymbol{r}'\, V_{ee}(\boldsymbol{r}-\boldsymbol{r}')\rho(\boldsymbol{r}') + \frac{\delta E_{\text{xc}}[\rho]}{\delta\rho(\boldsymbol{r})}$$
Solve Kohn-Sham equations [Eq. (1.8)] $$\left[-\frac{\hbar^2}{2m}\nabla^2 + V_{\text{eff}}(\boldsymbol{r}) - \varepsilon_i\right]\varphi_i(\boldsymbol{r}) = 0$$
Calculate electronic density [Eq. (1.6)], $$\rho(\boldsymbol{r}) = \sum_i^N \lvert\varphi_i(\boldsymbol{r})\rvert^2$$
Iterate to self-consistency
Calculate band structure $\varepsilon_i(\boldsymbol{k})$ [Eq. (1.8)], partial and total DOS, self-consistent Hamiltonian [Eq. (1.19)] $\Longrightarrow$ LDA+DMFT, total energy $E[\rho]$ [Eq. (1.4)], ...

Figure 1.2: Flow diagram of the DFT/LDA calculations, from Held et al. (2003).

respectively. Gunnarsson and Lundqvist (1976) observed that due to the isotropy of the Coulomb correlation, the exchange correlation energy depends only on the spherical average of $\rho_{\text{xc}}(\boldsymbol{r}, \boldsymbol{r}'-\boldsymbol{r})$. Furthermore, they found that if the sum rule

$$\int d\boldsymbol{r}'\rho_{\text{xc}}(\boldsymbol{r}, \boldsymbol{r}'-\boldsymbol{r}) = -1 \tag{1.14}$$

is satisfied, $E_{\text{xc}}[\rho]$ depends only weakly on the details of $\rho_{\text{xc}}(\boldsymbol{r}, \boldsymbol{r}'-\boldsymbol{r})$.

If the exchange correlation term $E_{\text{xc}}[\rho]$ would be known exactly, the DFT/LDA calculation could be done self-consistently without any approximations as shown in the flow diagram of Fig. 1.2.

1.1.2 Local density approximation

The derivations of the previous section were done without any approximations, the difficulty of the many-body problem was only transfered to the unknown exchange correlation functional $E_{\text{xc}}[\rho]$ (or to the exchange correlation hole $\rho_{\text{xc}}(\boldsymbol{r},\boldsymbol{r}'-\boldsymbol{r})$, respectively). In order to solve the DFT equations, this functional has to be approximated. One of the simplest approximations that has turned out to be unexpectedly successful is the local density approximation (LDA). Here, the full pair correlation function of Eq. (1.13) of the many-body system is at each point in space replaced by the pair correlation function of the homogeneous electron gas $g_0(\boldsymbol{r},\boldsymbol{r}',\lambda,\rho(\boldsymbol{r}))$ with local charge density $\rho(\boldsymbol{r})$, leading to

$$\rho_{\text{xc}}(\boldsymbol{r},\boldsymbol{r}'-\boldsymbol{r}) \equiv \rho(\boldsymbol{r}') \int_0^1 d\lambda [g_0(\boldsymbol{r},\boldsymbol{r}',\lambda,\rho(\boldsymbol{r})) - 1]. \tag{1.15}$$

This approximation satisfies the sum rule Eq. (1.14) which is one of the main reasons for its wide applicability. Inserting this form of $\rho_{\text{xc}}(\boldsymbol{r},\boldsymbol{r}'-\boldsymbol{r})$ into Eq. (1.12), one obtains

$$E_{\text{xc}}[\rho] \to \int d^3r \, \varepsilon_{\text{xc}}^{\text{LDA}}(\rho(\boldsymbol{r})), \tag{1.16}$$

where $\varepsilon_{\text{xc}}^{\text{LDA}}(\rho(\boldsymbol{r}))$ is the exchange and correlation energy of the homogeneous electron gas with density $\rho(\boldsymbol{r})$. It can be calculated from the perturbative solution (Hedin and Lundqvist 1971, von Barth and Hedin 1972) or numerical simulations (Ceperley and Alder 1980) of the jellium model which has a constant ionic background ($V_{ion}(\boldsymbol{r}) = const.$).

Although DFT/LDA is in principle only valid for the calculation of static properties like the ground-state energy, one of its main applications is the band structure calculation. To accomplish this, the Lagrange parameters ε_i of Eq. (1.8) are interpreted as physical one-particle energies of the system. Since the true ground state of the system under consideration is not a simple one-particle wave function, this is a further approximation beyond DFT. The Hamiltonian (1.1) is replaced by

$$\begin{aligned} \hat{H}_{\text{LDA}} &= \sum_\sigma \int d^3r \, \hat{\Psi}^+(\boldsymbol{r},\sigma) \left[-\frac{\hbar^2}{2m_e}\Delta + V_{\text{ion}}(\boldsymbol{r}) + \int d^3r' \, \rho(\boldsymbol{r}') V_{\text{ee}}(\boldsymbol{r}-\boldsymbol{r}') \right. \\ &\qquad \left. + V_{\text{xc}}^{\text{LDA}}(\boldsymbol{r}) \right] \hat{\Psi}(\boldsymbol{r},\sigma) \end{aligned} \tag{1.17}$$

with $V_{\text{xc}}^{\text{LDA}}(\boldsymbol{r}) = \frac{\delta E_{\text{xc}}^{\text{LDA}}[\rho]}{\delta\rho(\boldsymbol{r})}$. For calculation purposes, the field operators $\hat{\Psi}^+(\boldsymbol{r},\sigma)$ have to be expanded in a suitable set of basis functions Φ_{ilm}. There exists a multitude of basis sets that are suitable for different applications, Capelle (2003) gives an overview. In solid-state physics, linear muffin-tin orbitals (LMTO) and linear augmented plane waves (LAPW) are widely used (Andersen 1975, Gunnarsson, Jepsen and Andersen 1983, Andersen and Jepsen 1984). With the field operators in LMTO basis (i is the lattice site, l and m are orbital indices),

$$\hat{\Psi}^{+}(\boldsymbol{r},\sigma) = \sum_{ilm} \hat{c}^{\sigma\dagger}_{ilm} \Phi_{ilm}(\boldsymbol{r}) \ , \tag{1.18}$$

the Hamiltonian (1.17) can be written as

$$\hat{H}_{\mathrm{LDA}} = \sum_{ilm,jl'm',\sigma} (\delta_{ilm,jl'm'}\, \varepsilon_{ilm}\, \hat{n}^{\sigma}_{ilm} + t_{ilm,jl'm'}\, \hat{c}^{\sigma\dagger}_{ilm} \hat{c}^{\sigma}_{jl'm'}). \tag{1.19}$$

Here, $\hat{n}^{\sigma}_{ilm} = \hat{c}^{\sigma\dagger}_{ilm}\hat{c}^{\sigma}_{ilm}$,

$$t_{ilm,jl'm'} = \Big\langle \Phi_{ilm} \Big| -\frac{\hbar^2\Delta}{2m_e} + V_{\mathrm{ion}}(\boldsymbol{r}) + \int d^3r' \rho(\boldsymbol{r}') V_{\mathrm{ee}}(\boldsymbol{r}-\boldsymbol{r}') + \frac{\delta E^{\mathrm{LDA}}_{\mathrm{xc}}[\rho]}{\delta\rho(\boldsymbol{r})} \Big| \Phi_{jl'm'} \Big\rangle \tag{1.20}$$

for $ilm \neq jl'm'$ and zero otherwise; ε_{ilm} denotes the corresponding diagonal part.

The self-consistent solution of the Hamiltonian (1.19) together with the calculation of the electronic density (1.6) within the DFT/LDA, as shown in the flow diagram (Fig. 1.2), has been highly successful for the calculation of static properties and band structures for weakly correlated materials (Jones and Gunnarsson 1989). However, since correlations are only treated on a very rudimentary level, LDA fails to describe strongly correlated systems.

1.1.3 LDA and local Coulomb correlations

In the LDA, the only correlation that is implicitly taken into account is the Hartree term

$$E_{\mathrm{H}} = \frac{1}{2} \int d^3r \int d^3r' V_{\mathrm{ee}}(\boldsymbol{r}-\boldsymbol{r}')\rho(\boldsymbol{r})\rho(\boldsymbol{r}'). \tag{1.21}$$

For large coordination number, the nearest-neighbor density-density interaction, which is the largest non-local contribution, reduces to the Hartree term (Müller-Hartmann 1989, Wahle, Blümer, Schlipf, Held and Vollhardt 1998). The largest local contribution is the Coulomb interaction between electrons of the d- and f-orbitals on the same lattice site. These localized orbitals have an extensive overlap with respect to the Coulomb interaction leading to strong correlations in materials with partly filled d- or f-shells. Such local Coulomb interactions are not properly taken into account in LDA, but one can include them by supplementing the LDA Hamiltonian (1.19) with the most important matrix elements of the local Coulomb matrix:

$$\begin{aligned}\hat{H} &= \hat{H}_{\text{LDA}} - \hat{H}^{U}_{\text{LDA}} + \frac{1}{2}\sum_{i=i_d,l=l_d}\sum_{m\sigma,m'\sigma'}{}' U^{\sigma\sigma'}_{mm'}\hat{n}_{ilm\sigma}\hat{n}_{ilm'\sigma'} \\ &-\frac{1}{2}\sum_{i=i_d,l=l_d}\sum_{m\sigma,m'}{}' J_{mm'}\hat{c}^{\dagger}_{ilm\sigma}\hat{c}^{\dagger}_{ilm'\bar{\sigma}}\hat{c}_{ilm'\sigma}\hat{c}_{ilm\bar{\sigma}}. \qquad (1.22)\end{aligned}$$

Here, $U^{\sigma\sigma'}_{mm'}$ is the Coulomb repulsion and the z-component of the Hund's rule coupling and $J_{mm'}$ is the spin-flip term of the Hund's rule coupling between the localized electrons (we assume that only the bands of the atoms with index d are correlated, i.e., $i = i_d$ and $l = l_d$).[7] The prime on the sum indicates that at least two of the indices of an operator have to be different, and $\bar{\sigma} =\downarrow (\uparrow)$ for $\sigma =\uparrow (\downarrow)$. In the calculations presented in chapters 3, 4 and 5, the interaction parameters are simplified to $U^{\uparrow\downarrow}_{mm} \equiv U$, $J_{mm'} \equiv J$ and $U^{\sigma\sigma'}_{mm'} = U - 2J - J\delta_{\sigma\sigma'}$ for $m \neq m'$.[8] Thus, the interaction between different orbitals is $U' = U - 2J$ for electrons with different spin and $U' - J = U - 3J$ for spin-aligned electrons. The average Coulomb interaction $\bar{U}$ for M interacting orbitals is defined as[9]

$$\bar{U} = \frac{U + (M-1)(U-2J) + (M-1)(U-3J)}{2M-1}. \qquad (1.23)$$

Since the d-bands are assumed to be correlated, the one-particle potential of those orbitals will be shifted by the Coulomb interaction with respect to the s- and p-orbitals also included in the LDA Hamiltonian. Furthermore, the LDA already contains correlations in the form of the Hartree term (1.21). To account for the energy shift and to avoid the double counting of the correlations already included in LDA, the term $\hat{H}^{U}_{\text{LDA}}$ is subtracted in the Hamiltonian (1.22). As there does not exist a diagrammatic link between the LDA and the model Hamiltonian approach, this contribution cannot be rigorously expressed in terms of U and ρ. How it can be approximated numerically is demonstrated in Sec. 1.1.4.

Combining this last term with the LDA Hamiltonian, one gets a Hamiltonian $\hat{H}^{0}_{\text{LDA}}$ which describes the LDA one-particle energies without the local Coulomb interaction:

$$\hat{H}^{0}_{\text{LDA}} = \sum_{ilm,jl'm',\sigma}(\delta_{ilm,jl'm'}\,\varepsilon^{0}_{ilm}\;\hat{n}^{\sigma}_{ilm} + t_{ilm,jl'm'}\;\hat{c}^{\sigma\dagger}_{ilm}\hat{c}^{\sigma}_{jl'm'}). \qquad (1.24)$$

[7] In our calculations, this includes only the three t_{2g}-bands, but one can also include all five d-bands or the seven f-bands.

[8] The term $2J$ is due to the reduced Coulomb repulsion between different orbitals, the second term $J\delta_{\sigma\sigma'}$ arises from the z-component of the Hund's rule coupling.

[9] For three orbitals, the average LDA Coulomb interaction $\bar{U}$ is equal to the inter-orbital Coulomb interaction U', see Held et al. (2002) and Zölfl, Pruschke, Keller, Poteryaev, Nekrasov and Anisimov (2000) for details.

ε^0_{ilm} is equal to ε_{ilm} of Eq. (1.19) for the non-interacting orbitals and $\varepsilon^0_{il_dm} = \varepsilon_{il_dm} - \Delta\epsilon_d$ for the interacting orbitals. In reciprocal space, the matrix elements of $\hat{H}^0_{\rm LDA}$ are given by

$$(H^0_{\rm LDA}(\boldsymbol{k}))_{qlm,q'l'm'} = (H_{\rm LDA}(\boldsymbol{k}))_{qlm,q'l'm'} - \delta_{qlm,q'l'm'}\delta_{ql,q_dl_d}\Delta\epsilon_d n_d. \quad (1.25)$$

where q is an index of the atom in the unit cell, q_d denotes the atoms with interacting orbitals in the unit cell and $(H_{\rm LDA}(\boldsymbol{k}))_{qlm,q'l'm'}$ is the matrix element of (1.19) in $\boldsymbol{k}$-space. Eq. (1.22) can now be rewritten as the sum of the non-interacting part $\hat{H}^0_{\rm LDA}$ and the local Coulomb interaction:

$$\begin{aligned}\hat{H} &= \hat{H}^0_{\rm LDA} + \frac{1}{2}\sum_{i=i_d,l=l_d}\sum_{m\sigma,m'\sigma'}{}' U^{\sigma\sigma'}_{mm'}\hat{n}_{ilm\sigma}\hat{n}_{ilm'\sigma'} \\ &-\frac{1}{2}\sum_{i=i_d,l=l_d}\sum_{m\sigma,m'}{}' J_{mm'}\hat{c}^\dagger_{ilm\sigma}\hat{c}^\dagger_{ilm'\bar{\sigma}}\hat{c}_{ilm'\sigma}\hat{c}_{ilm\bar{\sigma}}. \end{aligned} \quad (1.26)$$

Here, $\hat{H}$ is the approximated *ab initio* Hamiltonian for the material under investigation and is the basis for the subsequent calculations. It was proposed by Anisimov (Anisimov et al. 1991, Anisimov, Aryasetiawan and Lichtenstein 1997) who obtained it in their LDA+U approach. Since the LDA+U does not contain true many-body physics (interactions are only treated in Hartree approximation), it cannot describe strongly correlated paramagnetic systems (see the introductory remarks of chapter 1). Various schemes that go beyond LDA+U and better approximate the full electron-electron interaction have been proposed and applied to different systems (Anisimov, Poteryaev, Korotin, Anokhin and Kotliar 1997, Lichtenstein and Katsnelson 1998, Drchal, Janiš and Kudrnovský 1999, Lægsgaard and Svane 1999, Wolenski 1999, Zölfl et al. 2000). One of the most successful approaches is the solution of the Hamiltonian (1.26) in DMFT (so-called LDA+DMFT) which was first implemented by Anisimov, Poteryaev, Korotin, Anokhin and Kotliar (1997) (for details on DMFT, see Sec. 1.2.2).

1.1.4 Constrained LDA

In model calculations, the local Coulomb interaction U and also the Hund's rule exchange constant J are parameters which are adjusted to fit to experiments or to yield certain physical properties, e.g., a metal-insulator transition. In order to obtain an *ab initio* LDA+DMFT scheme, those parameters have to be determined, too. This obviously is a difficult task and makes further approximations necessary. Within the constrained LDA method (McMahan, Martin and Satpathy 1988, Gunnarsson, Andersen, Jepsen and Zaanen 1989), it is possible to calculate those interaction parameters and also take into account screening

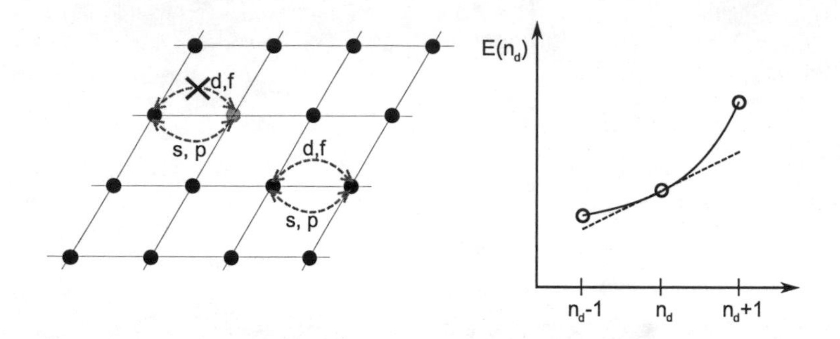

Figure 1.3: Left: In constrained LDA, the hopping matrix elements of the interacting d- or f-electrons on one lattice site (filled grey circle) are set to zero, only the non-interacting s- and p-electrons can still hop to and from this site and participate in the screening (dashed arrows). Right: When the number n_d of interacting d- or f-electrons on the decoupled site is changed, the LDA energy $E(n_d)$ changes non-linearly due to the correlations (circles); the straight line sketches the behavior of $\hat{H}_{\text{LDA}}$ (1.19).

effects. In constrained LDA, a LDA calculation is done for the system of interest with the interacting electrons (t_{2g}-, d- or f-electrons) confined to their site, i.e., their hopping matrix elements $t_{il_dm,jlm'}$ set to zero. The number of interacting electrons on this decoupled site can then be changed artificially, the other electrons (s- and p-electrons) can still hop and are redistributing to screen the decoupled, interacting electrons (Fig. 1.3). Starting from the number of interacting electrons n per site obtained in a standard LDA calculation (which normally is close to the number of electrons n_d inferred from the oxidation state,[10] e.g., two electrons per site for V_2O_3), one can perform constrained LDA calculations for $n_d - 1$, n_d and $n_d + 1$ electrons on the decoupled site and obtain the total energy as a function of the orbital occupation as shown in Fig. 1.3. With the Hamiltonian $\hat{H}_{\text{LDA}}$ of Eq. (1.19), the energy would change linearly with the number of interacting electrons:

$$E(n_d) = E_0 + \epsilon_d^{\text{LDA}}\, n_d \tag{1.27}$$

with $\epsilon_d^{\text{LDA}} = dE^{\text{LDA}}/dn_d$ (straight line in Fig. 1.3). However, in the interacting system (with the Hamiltonian of Eq. (1.22)), the energy required to add the n_dth electron is lower than the energy required to add the $(n_d + 1)$th electron:[11]

$$E(n_d) = E_0 + (1/2)\,\bar{U} n_d(n_d - 1) + (\epsilon_d^{\text{LDA}} + \Delta\epsilon_d)n_d. \tag{1.28}$$

[10] This number of electrons n_d is also used in the subsequent LDA+DMFT calculation.

[11] Note that the Coulomb contribution $(1/2)\, d[\bar{U} n_d(n_d - 1)]/dn_d = U(n_d - 1/2)$ is cancelled out by one part of $\Delta\epsilon_d$.

This leads to the curvature of $E(n_d)$ in Fig. 1.3 for the interacting system compared to the non-interacting system. Since the interacting system is described by the Hamiltonian (1.26), the LDA energies obtained for the three different numbers of interacting particles on the decoupled lattice site have to be reproduced with this Hamiltonian. Thus, $\bar{U}$ and $\Delta\epsilon_d$ can be determined by fitting these parameters to reproduce the constrained LDA energies. The Hund's exchange parameter J can be obtained in the same way in a constrained LDA calculation with spin polarization.

Although constrained LDA is presently the best method to obtain the Coulomb interaction parameters, a significant uncertainty remains, especially for the parameter U. While the total LDA spectrum has only a small dependence on the choice of the basis set, the shape of the orbitals which are considered to be interacting has a strong influence on the result of the constrained LDA calculations of U, even when an appropriate basis like the LMTO is used.[12] For example, two calculations for $LaTiO_3$ (with a Wigner-Seitz radius of 2.37 a.u. for Ti), a LMTO-ASA calculation by Nekrasov, Held, Blümer, Poteryaev, Anisimov and Vollhardt (2000)[13] and a ASA-LMTO calculation within orthogonal representation by Solovyev, Hamada and Terakura (1996) yielded $U = 4.2$ eV and $U = 3.2$ eV, respectively. In cases where a precise value of U is crucial for the physical properties (e.g., for the metal-insulator transition in V_2O_3, see chapter 3), the U from constrained LDA can only be a starting point and has to be adjusted to correctly describe the correlated system. In other cases (e.g., for $SrVO_3$ and $CaVO_3$, see chapter 4), the Coulomb parameters from constrained LDA correctly describe the correlated system and no adjustments are necessary to obtain the correct physics and good agreement with experiments.

1.2 Many-body theory

In many-body theory, instead of making approximations and trying to solve the Hamiltonian (1.1) directly as in DFT, simplified, abstract models are formulated. Those models incorporate certain characteristic physics of real systems, but offer only a limited insight due to their inherent simplicity. In some cases, the models can be solved analytically, but often only numerical solutions are possible. One of the simplest models for describing strongly correlated systems is the Hubbard model (Gutzwiller 1963, Hubbard 1963, Kanamori 1963) which will be disscussed in the following.

[12] The Hund's exchange parameter J is less affected by the choice of the basis since it only has a small dependence on the screening.

[13] In this calculation, the TB LMTO-ASA code by (Andersen 1975, Gunnarsson et al. 1983, Andersen and Jepsen 1984) was used.

1.2.1 The Hubbard model

In the Hubbard model, the complicated electron-electron interaction term (1.3), which describes the interplay of each electron with every other electron, is replaced by a purely local term, i.e., an interaction only between the electrons on the same lattice site. Furthermore, only isotropic hopping to nearest neighbors is allowed. Thus, the Hamiltonian of the one-band Hubbard model can be written as

$$\hat{H} = -t \sum_{<ij>,\sigma} \hat{c}^{\dagger}_{i\sigma} \hat{c}_{j\sigma} + U \sum_{i} \hat{n_{i\uparrow}} \hat{n_{i\downarrow}}. \tag{1.29}$$

Here, the sum is restricted to nearest neighbor pairs i, j (indicated by the brackets) and the kinetic energy (hopping) and the Coulomb interaction are parametrized by t and U, respectively. Since the systems under investigation in this work are V-3d systems with 5 d-bands (3 t_{2g}-bands and 2 e_g^{σ}-bands), a one-band model is not sufficient to describe those materials. Especially the Hund's rule coupling that energetically favors aligned spins is important, see chapter 3 for details. Taking those interactions into account, one obtains a multi-band Hubbard model. Its Hamiltonian can be written as[14]

$$\begin{aligned} \hat{H} &= -t \sum_{<ij>,m\sigma} \hat{c}^{\dagger}_{im\sigma} \hat{c}_{jm\sigma} + \frac{1}{2} \sum_{i} \sum_{m\sigma,m'\sigma'}{}' U^{\sigma\sigma'}_{mm'} \hat{n}_{im\sigma} \hat{n}_{im'\sigma'} \\ &- \frac{1}{2} \sum_{i} \sum_{m\sigma,m'}{}' J_{mm'} \hat{c}^{\dagger}_{im\sigma} \hat{c}^{\dagger}_{im'\bar{\sigma}} \hat{c}_{im'\sigma} \hat{c}_{im\bar{\sigma}}. \end{aligned} \tag{1.30}$$

The prime on the sum indicates that at least two of the indices of an operator have to be different,[15] and $\bar{\sigma} = \downarrow (\uparrow)$ for $\sigma = \uparrow (\downarrow)$. In quantum Monte Carlo simulations, the spin-flip term $J_{mm'}$ in general leads to a serious numerical (minus-sign) problem (Held 1999).[16] We therefore take into account only the z-component of the Hund's rule coupling which is included in $U^{\sigma\sigma'}_{mm'}$ and neglect the spin-flip term.[17] As disscussed in Sec. 1.1.3, the interaction parameters can be simplified so that one retains only density-density terms, namely the intra-band Coulomb interaction U, the inter-band interaction U' and the Hund's rule coupling J. Replacing the simple hopping term with parameter t by a more general term with

[14] Note that an analogous interaction term was introduced in Eq. (1.22) of Sec. 1.1.3.

[15] In the literature, the sum is often written as $\sum_{m<m'}$ which is equivalent to $\frac{1}{2}\sum'_{mm'}$.

[16] For the particle-hole symmetric case, the inclusion of a spin-flip term without a minus-sign problem is possible (Motome and Imada 1997). Recently, a new scheme was proposed which reduces the minus-sign problem also away from half-filling (Sakai, Arita and Aoki 2004) for small interaction strengths. Another promising new approach in that respect is the continuous-time QMC (Rubtsov, Savkin and Lichtenstein 2004).

[17] This is equivalent to replacing the full Heisenberg interaction by an Ising interaction which breaks the SU(2) symmmetry of the problem (Motome and Imada 1997, Held and Vollhardt 1998) but retains much of the physics of the Hund's rule coupling.

the dispersion relation $\epsilon_{\boldsymbol{k}m}$, one obtains the multi-band Hubbard Hamiltonian

$$\hat{H} = \sum_{\boldsymbol{k},m\sigma} \epsilon_{\boldsymbol{k}m} \hat{c}^{\dagger}_{\boldsymbol{k}m\sigma} \hat{c}_{\boldsymbol{k}m\sigma} + U \sum_{im} \hat{n}_{im\uparrow} \hat{n}_{im\downarrow} + \frac{1}{2} \sum_{i,mm',\sigma\sigma'}{}' (U' - \delta_{\sigma\sigma'} J) \hat{n}_{im\sigma} \hat{n}_{im'\sigma'}. \quad (1.31)$$

Since even in the one-band Hubbard model (1.29), the interaction and hopping terms do not commute, a simultaneous diagonalization is not possible, making the solution of the Hubbard model highly non-trivial. In dimension $d = 1$ an exact solution of the Hubbard model was found using the Bethe ansatz (Bethe 1931, Lieb and Wu 1968). In higher dimensions, exact statements can only be made for special cases.[18] Therefore, additional approximations are necessary, especially for extensions to the simple Hubbard model like the multi-band case described above. One of the most successful approximations in this respect is the DMFT which will be described in the next section.

1.2.2 Dynamical mean-field theory

In mean-field theory, the complicated problem of one site interacting with all other sites is replaced with the problem of one site coupled to an effective mean field. Thus, the original many-body problem is reduced to an effective one-body problem. In three-dimensional systems, this is only an approximation that is controlled by the parameter $1/Z$. The coordination number Z, i.e., the number of nearest neighbors, typically is of the order 10 in three-dimensional lattices, e.g., for simple cubic, body-centered cubic and face-centered cubic lattices, $Z = 6, 8, 12$, respectively. For the hypercubic lattice, which is the generalization of the cubic lattice to arbitrary dimension d, there is a simple relation between the coordination number and the dimension, $Z = 2d$. In the limit $Z \to \infty$ (which is equal to the *limit of infinite dimensions* $d \to \infty$), the mean-field theory is exact. One well-known example for such a mean-field theory is the Weiss mean-field approximation (Weiss 1907) for the Heisenberg model

$$\hat{H}_{Heisenberg} = -J \sum_{<ij>} \hat{\boldsymbol{S}}_i \hat{\boldsymbol{S}}_j \quad (1.32)$$

In order to get a non-trivial energy per site in the limit $Z \to \infty$, J has to be rescaled, $J = J^*/Z$. In the limit of infinite coordination number, the central limit theorem holds and fluctuations between neighboring lattice sites vanish. One can then use the Hartree decoupling scheme $\hat{\boldsymbol{S}}_i \hat{\boldsymbol{S}}_j \to \hat{\boldsymbol{S}}_i \langle \hat{\boldsymbol{S}}_j \rangle + \langle \hat{\boldsymbol{S}}_i \rangle \hat{\boldsymbol{S}}_j - \langle \hat{\boldsymbol{S}}_i \rangle \langle \hat{\boldsymbol{S}}_j \rangle$. Choosing the z-axis as the quantization axis, the mean-field Hamiltonian

$$\hat{H}_{MF} = -2J^* \sum_i \hat{S}^z_i h_i + J^* \sum_i \langle \hat{S}^z_i \rangle h_i, \quad (1.33)$$

[18] One example is the Nagaoka (1966) theorem which states that for $U/t \to \infty$ at half filling plus one electron, the ground state of the Hubbard model is ferromagnetic.

which is identical to the mean-field Hamiltonian for the Ising model, is obtained. The mean-field h_i is the averaged spin of all nearest-neighbor (NN) sites, $h_i = \frac{1}{Z}\sum_{j \text{ NN of } i}\langle \hat{S}_i^z \rangle$. The mean-field equations have to be solved self-consistently. In the limit $Z \to \infty$, this Weiss mean-field approximation becomes exact (Brout 1960).

A similar mean-field theory for itinerant correlated fermions (e.g., the Hubbard model (1.29)) can be formulated, but there are some important differences that have to be considered. The Hartree-Fock decoupling of the local density-density interaction (the analog to the Hartree decoupling used for the Heisenberg model above) is not exact in the limit $Z \to \infty$, i.e., a static mean-field is not appropriate in such systems. In order to find a nontrivial $Z \to \infty$ limit for the Hubbard model, its parameters have to be rescaled to conserve the interplay of kinetic and potential energy. Since the Coulomb interaction is purely local and therefore independent of coordination number, only the hopping parameter t has to be rescaled. Metzner and Vollhardt (1989) showed that for the Hubbard model on a hypercubic lattice with only nearest neighbor hopping, the scaling

$$t = \frac{t^*}{\sqrt{Z}} \tag{1.34}$$

ensures a finite and non-zero density of states (DOS). Metzner and Vollhardt (1989) and Müller-Hartmann (1989) found that for this particular scaling the self-energy becomes purely local in the limit of infinite coordination number. For generalized lattices, Müller-Hartmann (1989) showed that the momentum conservation is irrelevant in the limit of infinite dimensions. Furthermore, he found that of the density-density interactions only the local Coulomb interaction remains dynamical in infinite dimensions whereas all other interactions are reduced to their Hartree contribution. Due to the purely local self-energy

$$\Sigma^\sigma_{ij}(\omega) \xrightarrow{Z\to\infty} \delta_{ij}\Sigma^\sigma(\omega) \qquad \text{or} \qquad \Sigma^\sigma(\boldsymbol{k},\omega) \xrightarrow{Z\to\infty} \Sigma^\sigma(\omega), \tag{1.35}$$

which retains its frequency dependence, the local lattice Green function $G^\sigma(\omega)$ for the Hubbard model (1.29) can be written as

$$G^\sigma(\omega) = \frac{1}{V_B}\sum_{\boldsymbol{k}} \frac{1}{\omega + \mu - \epsilon_{\boldsymbol{k}} - \Sigma^\sigma(\omega)}, \tag{1.36}$$

with the volume of the Brillouin zone V_B, the chemical potential μ and the energy $\epsilon_{\boldsymbol{k}}$. The sum extends over all $\boldsymbol{k}$-points of the Brillouin zone.

For the LDA Hamiltonian with Coulomb interactions (1.26), a similar $\boldsymbol{k}$-integrated Dyson equation can be stated:

$$G^\sigma_{qlm,q'l'm'}(\omega) = \frac{1}{V_B}\int d^3k \, \left(\left[\omega\mathbb{1} + \mu\mathbb{1} - H^0_{\text{LDA}}(\boldsymbol{k}) - \Sigma^\sigma(\omega) \right]^{-1} \right)_{qlm,q'l'm'}. \tag{1.37}$$

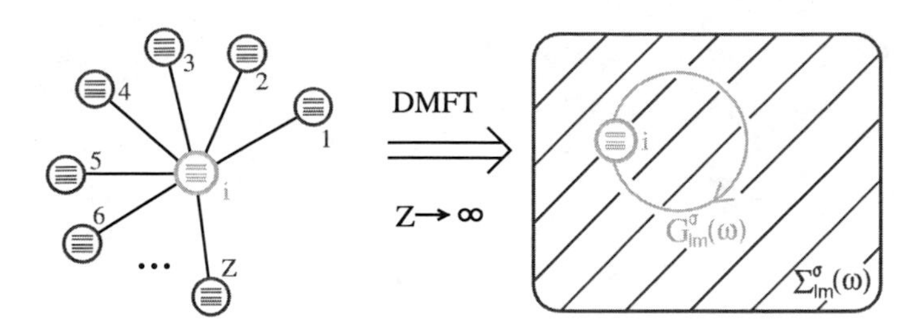

Figure 1.4: If the number of neighboring lattice sites Z goes to infinity, the influence of those neighboring sites can be replaced by a dynamical mean-field described by the (multi-band) self-energy $\Sigma^{\sigma}_{lm}(\omega)$. This DMFT problem is equivalent to the self-consistent solution of the $\boldsymbol{k}$-integrated Dyson equation (1.37) and the multi-band Anderson impurity model Eq. (1.39).

Here, $[...]^{-1}$ is the inversion of the matrix with elements (qlm) and $(q'l'm')$, $H^0_{\text{LDA}}(\boldsymbol{k})$ is the non-interacting LDA Hamiltonian defined in Eq. (1.25) and the self-energy $\Sigma^{\sigma}(\omega)$ is non-zero only for the interacting orbitals. For systems with well-separated and degenerate interacting bands, the Dyson equation can be further simplified, see Sec. 1.2.3 for details. It is important to note that this Dyson equation and also the DMFT formalism described here is only valid for systems with a high symmetry and thus diagonal Green functions and self-energies. For the transition metal oxides under investigation in this work, this poses no restriction, it is even possible to simplify the algorithm further as described in Sec. 1.2.3. For more complex materials with strong off-diagonal elements in the self-energy, the formalism has to be generalized (Lichtenstein, Katsnelson and Kotliar 2003).

The momentum independence of the self-energy and thus the lattice Green function has an important consequence for the Hubbard model. In the limit of infinite dimensions, the interaction of one site with all other lattice sites can be replaced with a dynamical (frequency dependent) mean-field, i.e., the Hubbard model can be mapped onto the self-consistent solution of a single impurity Anderson model (SIAM)[19](Georges and Kotliar 1992, Jarrell 1992, Janiš and Vollhardt 1992), see Fig. 1.4. The local non-interacting (bath) Green function $\mathcal{G}(\omega)$ of the Anderson model is connected to the local lattice Green function $G(\omega)$ of the Hubbard model through

$$\mathcal{G}(\omega)^{-1} = G(\omega)^{-1} + \Sigma(\omega). \tag{1.38}$$

In the imaginary-time domain,[20] the local one-particle Green function can be

[19] A summary on the history and the physics of this model can be found in Anderson (1972).
[20] The use of the formulation in imaginary time τ or in its Fourier transform, the fermionic Matsubara frequency ω_ν, is convenient since it is also used for the imaginary time quantum

written as a functional integral over the Grassmann variables ψ and ψ^*:[21]

$$G^{\sigma}_{\nu m} = -\frac{1}{\mathcal{Z}} \int \mathcal{D}[\psi]\mathcal{D}[\psi^*] \psi^{\sigma}_{\nu m} \psi^{\sigma *}_{\nu m} e^{\mathcal{A}[\psi,\psi^*,\mathcal{G}^{-1}]}. \tag{1.39}$$

Here, $G^{\sigma}_{\nu m} \equiv G^{\sigma}_{m}(i\omega_\nu)$ with the fermionic Matsubara frequency $i\omega_\nu = i(2\nu+1)\pi/\beta$ (β: inverse temperature) and orbital index m ($l = l_d$, $q = q_d$). The partition function is defined as $\mathcal{Z} = \int \mathcal{D}[\psi]\mathcal{D}[\psi^*] \exp(\mathcal{A}[\psi,\psi^*,\mathcal{G}^{-1}])$ and the single site action $\mathcal{A}$ in terms of the imaginary time τ is

$$\begin{aligned}\mathcal{A}[\psi,\psi^*,\mathcal{G}^{-1}] = & \sum_{\nu,\sigma,m} \psi^{\sigma *}_{\nu m} (\mathcal{G}^{\sigma}_{\nu m})^{-1} \psi^{\sigma}_{\nu m} \\ & -\frac{1}{2}\sum_{m\sigma,m\sigma'}{}' U^{\sigma\sigma'}_{mm'} \int_0^\beta d\tau\, \psi^{\sigma *}_{m}(\tau)\psi^{\sigma}_{m}(\tau)\psi^{\sigma' *}_{m'}(\tau)\psi^{\sigma'}_{m'}(\tau) \\ & +\frac{1}{2}\sum_{m\sigma,m}{}' J_{mm'} \int_0^\beta d\tau\, \psi^{\sigma *}_{m}(\tau)\psi^{\bar{\sigma}}_{m}(\tau)\psi^{\bar{\sigma} *}_{m'}(\tau)\psi^{\sigma}_{m'}(\tau) \ .\end{aligned} \tag{1.40}$$

The $\boldsymbol{k}$-integrated Dyson equation (1.37) and the single-site problem (1.39) that are coupled via (1.38) have to be solved self-consistently by an iterative procedure to obtain the DMFT solution. This is depicted in the flow diagram Fig. 1.5.

The solution of the Dyson equation (1.37) is rather straightforward, for simple model systems (e.g., with a semi-elliptical Bethe DOS (Georges et al. 1996)), it can even be done analytically. For realistic systems, a (numerically inexpensive) tetrahedron integration over the Brillouin zone (Jepson and Anderson 1971, Singhal 1975) has to be performed. In contrast, the solution of the single-site problem (1.39) is highly non-trivial. Fortunately, the equivalence to the Anderson impurity problem allows the use of most of the methods developed for the solution of the SIAM such as exact diagonalization (ED) (Caffarel and Krauth 1994, Georges et al. 1996), iterative perturbation theory (IPT) (Georges and Kotliar 1992, Georges et al. 1996), non-crossing approximation (NCA) (Keiter and Kimball 1970, Bickers, Cox and Wilkins 1987, Pruschke and Grewe 1989, Pruschke, Cox and Jarrell 1993), fluctuation exchange approximation (FLEX) (Bickers and Scalapino 1989, Bickers and White 1991), numerical renormalization group (NRG) (Wilson 1975, Bulla 2000), and quantum Monte Carlo (QMC) simulation. The QMC method will be presented in Sec. 2.1 and utilized for all calculations in this work. A comparison of the results of different methods (IPT, NCA, QMC) in LDA+DMFT can be found in Held et al. (2003).

Monte Carlo algorithm in Sec. 2.1.

[21] An introduction to Grassmann variables is given in Negele and Orland (1987) and in Berezin (1987).

Choose an initial self-energy Σ

Calculate G from Σ via the $\boldsymbol{k}$-integrated Dyson Eq. (1.37):

$$G^{\sigma}_{qlm,q'l'm'}(\omega) = \frac{1}{V_B} \int d^3k \left(\left[\omega \mathbb{1} + \mu \mathbb{1} - H^0_{\text{LDA}}(\boldsymbol{k}) - \Sigma^{\sigma}(\omega) \right]^{-1} \right)_{qlm,q'l'm'}$$

$\mathcal{G} = (G^{-1} + \Sigma)^{-1}$

Calculate G from $\mathcal{G}$ via the DMFT single-site problem Eq. (1.39)

$$G^{\sigma}_{\nu m} = -\frac{1}{\mathcal{Z}} \int \mathcal{D}[\psi]\mathcal{D}[\psi^*]\psi^{\sigma}_{\nu m}\psi^{\sigma *}_{\nu m} e^{\mathcal{A}[\psi,\psi^*,\mathcal{G}^{-1}]}$$

$\Sigma_{\text{new}} = \mathcal{G}^{-1} - G^{-1}$

Iterate with $\Sigma = \Sigma_{\text{new}}$ until convergence, i.e., $||\Sigma - \Sigma_{\text{new}}|| < \epsilon$

Figure 1.5: Flow diagram of the DMFT self-consistency cycle, from Held et al. (2003).

Due to its non-perturbative character and the diversity of available impurity solvers, the DMFT is very successful for the description of realistic (three-dimensional) strongly correlated systems. However, it fails for strongly momentum dependent phenomena like d-wave superconductivity and for problems where non-local correlations are important, e.g., the formation of a spin singlet on two neighboring sites. Attempts to systematically include $\mathcal{O}(1/d)$ corrections for finite-dimensional systems have been made by various authors (Gebhard 1990, Vlaming and Vollhardt 1992, van Dongen and Janiš 1994, Schiller and Ingersent 1995, Zaránd, Cox and Schiller 2000). Another approach to include non-local correlations is to treat a cluster of sites with DMFT instead of one single site. A short overview over these cluster approaches is given in Sec. 1.3.3.

1.2.3 Simplifications for well-separated and degenerate bands

Many transition metal oxides have a cubic crystal structure with a transition metal atom octahedrally coordinated by oxygen atoms. The electrons in the d orbitals of the transition metal in those systems are strongly correlated and have to be taken into account in the DMFT calculations. Due to the cubic crystal field of the oxygen, the d-orbitals are split into three degenerate t_{2g}-orbitals and two degenerate e_g^σ-orbitals, see, e.g., chapter 3 for details. Often, the splitting is sufficient to completely separate the t_{2g}- or e_g^σ-bands at the Fermi edge from the remaining bands (e.g., the O-2p bands). Thus, the low-energy physics of the system is well described by only including the partially filled bands at the Fermi edge in the DMFT calculations. The Green function and the self-energy of those bands simplify to $G_{qlm,q'l'm'}(\omega) = G_{mm'}(\omega)\delta_{ql,q'l'}$ and $\Sigma_{qlm,q'l'm'}(\omega) = \Sigma_{mm'}(\omega)\delta_{ql,q'l'}$ for $l = l_d$ and $q = q_d$ (where l_d and q_d denote the indices of the interacting bands at the Fermi energy). The Hamiltonian H^0_{LDA} (1.17) can be downfolded to a basis with only the interacting bands [22]. With this new Hamiltonian $H^{0\,\text{eff}}_{\text{LDA}}$ (and suppressing the indices $l = l_d$ and $q = q_d$), the Dyson equation (1.37) is simplified to

$$G_{mm'}(\omega) \;=\; \frac{1}{V_B}\int \mathrm{d}^3k \,\left(\left[\omega\mathbb{1} + \mu\mathbb{1} - H^{0\,\text{eff}}_{\text{LDA}}(\boldsymbol{k}) - \Sigma(\omega)\right]^{-1}\right)_{mm'}. \quad (1.41)$$

One possibility to get such a downfolded Hamiltonian is the Wannier function formalism described in Sec. 1.3.1, another method was proposed by Andersen, Saha-Dasgupta, Ezhov, Tsetseris, Jepsen, Tank, Arcangeli and Krier (2001).

If the interacting bands (l_d, q_d) are degenerate (as it is the case for, e.g., $SrVO_3$), the LDA Hamiltonian and the self-energy are diagonal ($(H^{0\,\text{eff}}_{\text{LDA}})_{mm'} = (H^{0\,\text{eff}}_{\text{LDA}})_m\delta_{mm'}$, $\Sigma_{mm'}(\omega) = \Sigma_m(\omega)\delta_{mm'}$) and the interacting Green function can be expressed via the non-interacting Green function $G^0_m(\omega)$:

$$G_m(\omega) \;=\; G^0_m(\omega - \Sigma(\omega)) = \int d\epsilon \frac{N^0_m(\epsilon)}{\omega - \Sigma_m(\omega) - \epsilon}. \quad (1.42)$$

Here, instead of a tetrahedron integration over the Brillouin zone as in Eq. (1.37), only an energy integration over the unperturbed LDA density of states $N^0_m(\epsilon)$ has to be performed, which makes the calculation considerably simpler (and numerically more stable). Furthermore, since the subtraction $\hat{H}^U_{\text{LDA}}$ in (1.42) only shifts the chemical potential, the exact form of $\hat{H}^U_{\text{LDA}}$ is not important for the simplified calculations. Another conceptual simplification is the applicability of Luttinger's theorem of Fermi pinning which states that in a Fermi liquid at $T = 0$, the interacting DOS at the Fermi energy is fixed at the value of the non-interacting DOS. This simplified LDA+DMFT approach is automatically self-consistent (in the sense of a fully self-consistent approach as described in Sec. 1.3.2) since the number of electrons in the interacting bands is fixed.

[22] In the downfolding procedure, the full LDA Hamiltonian is projected to a more atomic-like basis set and the interacting bands are extracted to yield a Hamiltonian that only includes those interacting bands.

The simplified Dyson equation (1.42) is strictly valid only when the interacting orbitals are degenerate and clearly separated from the rest of the orbitals. Anisimov, Kondakov, Kozhevnikov, Nekrasov, Pchelkina, Allen, Mo, Kim, Metcalf, Suga, Sekiyama, Keller, Leonov, Ren and Vollhardt (2005) showed in their comparative calculations that for a small overlap between the interacting orbitals and the remaining orbitals (as in $SrVO_3$) and for non-degenerate bands with nearly identical centers of gravity and bandwidth (as in V_2O_3), Eq. (1.42) is a justified approximation. All the calculations presented in this work for V_2O_3 (chapter 3), $Sr_{(1-x)}Ca_xVO_3$ (chapter 4) and LiV_2O_4 (chapter 5), are performed with this simplified scheme.

1.3 Extensions of the LDA+DMFT scheme

In this section, some interesting extensions of the LDA+DMFT scheme will be presented. Those include the Wannier function formalism for full Hamiltonian calculations, a fully self-consistent LDA+DMFT scheme and an overview over cluster extensions of the DMFT. Since the methods of this section are not directly employed in the calculations of this work, only a brief description with important references is given.

1.3.1 Wannier function formalism

The Wannier function formalism for LDA+DMFT calculations in full orbital space was proposed by Anisimov et al. (2005). In the LDA, the Hamiltonian is defined in the full orbital basis set $|\phi_\mu\rangle$ (e.g., in LMTO)[23]

$$\hat{H} = \sum_{\mu\nu} |\phi_\mu\rangle H_{\mu\nu} \langle\phi_\nu|. \tag{1.43}$$

The solution of the correlation problem in DMFT based on this full orbital LDA Hamiltonian is practically not possible due to the large number of orbitals and different interaction parameters for different types of bands. However, for low-energy properties, only the (partially filled) orbitals in the vicinity of the Fermi edge are important. By projecting the full orbital Hamiltonian to the subspace of partially filled orbitals, the computational complexity can be reduced and the correlation problem can be solved explicitly. For this projection, the choice of the basis set is of considerable importance. The orbitals of the basis set have to correctly describe the partially filled correlated bands while retaining the localized, site-centered atomic-like form. Therefore, the LMTOs or augmented spherical

[23] Greek indices μ and ν are used for matrices in full orbital space, latin indices n and n' for matrices in the subspace of interacting bands.

waves (ASWs) that are used for the LDA calculation are not appropriate.[24] A natural choice that fulfills those requirements are the Wannier functions (WFs) $|W_n^{\boldsymbol{T}}\rangle$ (with band index n and lattice translation vector $\boldsymbol{T}$). In the basis of Wannier functions, the Hamiltonian on the reduced subspace can be written as

$$\hat{H}^{WF} = \sum_{nn'\boldsymbol{T}} |W_n^{\boldsymbol{0}}\rangle H_{nn'}(\boldsymbol{T})\langle W_{n'}^{\boldsymbol{T}}|, \tag{1.44}$$

where the block matrix $H_{nn'}$ is a projection of the full orbital Hamilton operator (1.43) onto the subspace defined by the WFs. Thus, the total Hilbert space is divided into a direct sum of this subspace of correlated orbitals and a subspace orthogonal to it that consists of all other states (from the LDA calculation). The Coulomb interaction parameters can be calculated in a constrained LDA calculation (see Sec. 1.1.4) for the specific WF basis set for this subspace. With the projected Wannier Hamiltonian $\hat{H}^{WF}$ and the Coulomb interaction parameters from constrained LDA, the DMFT calculation can be performed as described in Sec. 1.2.2. The result from such a DMFT calculation is a local self-energy operator $\widehat{\Sigma}^{WF}(i\omega)$ that is defined in the basis of WFs centered on one site

$$\widehat{\Sigma}^{WF}(i\omega) = \sum_{nn'} |W_n^{\boldsymbol{0}}\rangle \Sigma_{nn'}(i\omega)\langle W_{n'}^{\boldsymbol{0}}|. \tag{1.45}$$

The self-energy matrix $\Sigma_{nn'}(i\omega)$ on the reduced subspace can then be converted back to a self-energy on the full Hilbert space $\Sigma_{\mu\nu}(i\omega)$, since the projection matrix between the full orbital basis set and the reduced WF basis retains all information on the orbitals below and above the correlated (projected) orbitals. Thus one can calculate (with the maximum entropy method) a full orbital Green function $G(\boldsymbol{r},\boldsymbol{r}',\omega)$ and spectral function $A(\omega)$, which contains the full information about the system and combines the weakly interacting orbitals (s- and p-orbitals) calculated with LDA and the strongly correlated orbitals (e.g., d-orbitals) computed with LDA+DMFT in a well-defined manner. The transformation back to full orbital space is a major advantage of the Wannier function formalism. As described in the next section, it allows one to reinsert the DMFT results into the LDA calculation and thus perform a fully self-consistent LDA+DMFT calculation.

Calculations within this Wannier function formalism for $SrVO_3$ and V_2O_3 were presented in Anisimov et al. (2005) and found to agree well with the results for the simplified LDA+DMFT scheme from Sekiyama, Fujiwara, Imada, Suga, Eisaki, Uchida, Takegahara, Harima, Saitoh, Nekrasov, Keller, Kondakov, Kozhevnikov, Pruschke, Held, Vollhardt and Anisimov (2004) and Keller, Held, Eyert, Anisimov and Vollhardt (2004) (which are also part of this work).

[24] For f-electron systems, LMTOs are a reasonable choice, but not for d-electron systems since the d-orbitals in LMTO are rather extended and hybridize strongly with the p-orbitals.

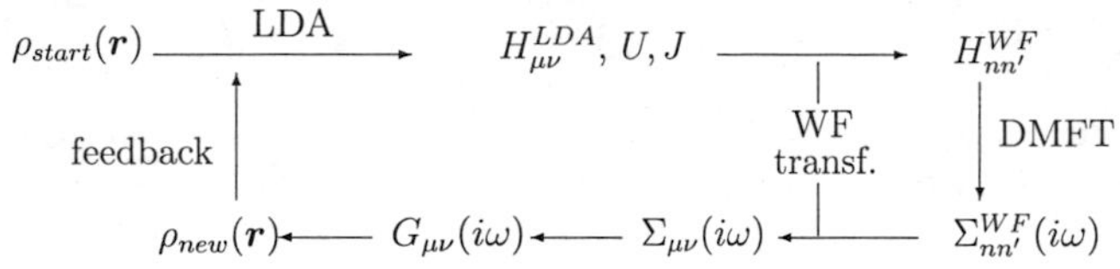

$\rho_{start}(\boldsymbol{r})$ —LDA→ $H_{\mu\nu}^{LDA}, U, J$ —WF transf.→ $H_{nn'}^{WF}$

feedback ↑ DMFT ↓

$\rho_{new}(\boldsymbol{r})$ ← $G_{\mu\nu}(i\omega)$ ← $\Sigma_{\mu\nu}(i\omega)$ ← $\Sigma_{nn'}^{WF}(i\omega)$

Figure 1.6: Self-consistent LDA+DMFT scheme within the Wannier function formalism.

1.3.2 Self-consistent LDA+DMFT calculations

In the LDA+DMFT procedure presented in this chapter so far, the result from the band structure calculations (the interaction parameters U and J and either a full or reduced LDA Hamiltonian or the non-interacting density of states) is only a static input into the DMFT self-consistent loop without any feedback to the LDA. However, the DMFT calculation will generally change the occupation of the correlated bands. This will cause a change of the electron density $\rho(\boldsymbol{r})$ and thus also a change of the LDA Hamiltonian $\hat{H}_{LDA}$ that depends on the electron density. Furthermore, the Coulomb interaction parameters are affected by the change of $\rho(\boldsymbol{r})$ and have to be recalculated in constrained LDA. It would be advantageous to use this new data that is modified by the correlations for a further DMFT calculation and thus get a feedback loop that is iterated until convergence is reached. Such a self-consistent LDA+DMFT scheme in the context of the Wannier function formalism is shown in Fig. 1.3.2. In this approach, the full orbital LDA Hamiltonian is projected to the subspace of interacting orbitals defined in WF basis, a DMFT calculation is performed and the self-energy resulting from this calculation is then transformed back from the subspace in WF basis to full orbital space (more details can be found in Anisimov et al. (2005)). In the self-consistency loop, not only the Coulomb interaction parameters U and J and the LDA Hamiltonian, but also the Wannier function basis used for the projection of the Hamiltonian is recalculated self-consistently. If the Coulomb interaction parameters are set to zero, this scheme reduces to the self-consistent solution of the Kohn-Sham equations as in a common LDA calculation. The impact of the self-consistency loop on the results of the LDA+DMFT calculation depends on the change of orbital occupation due to the correlations and thus on the particular system under investigation. For example, in the case of V_2O_3 (see chapter 3), the a_{1g} occupation found in LDA is halved for the DMFT calculation near the metal-insulator transition.

A self-consistent LDA+DMFT scheme similar to the one presented here (but without self-consistent calculation of U and in LMTO instead of Wannier function basis) was utilized by Savrasov and Kotliar (2001) in their calculations for plutonium.

1.3.3 Cluster extensions of DMFT

Although DMFT is a reliable and powerful method to describe strongly correlated systems with a local, $\boldsymbol{k}$-independent self-energy, it cannot describe phenomena with strong momentum dependence, e.g., d-wave superconductivity or spin-singlet formation on two neighboring sites. In DMFT, only a single site with local correlations in a dynamical mean-field is considered, non-local correlations are not taken into account. However, it is possible to extend the DMFT to treat not only a single site, but a small cluster of sites in a mean environment. A natural choice for such a cluster in the context of LDA+DMFT would be the correlated atoms in the unit cell of the material under investigation (e.g., for V_2O_3, the four vanadium atoms in the unit cell). This idea is the basis for the cluster extensions that will be sketched in the following, the dynamical cluster approximation (DCA) and the cluster (or cellular) DMFT (CDMFT). In the DCA (Hettler, Tahvildar-Zadeh and Jarrell 1998, Hettler, Mukherjee, Jarrell and Krishnamurthy 2000), see also (Jarrell, Maier, Hettler and Tahvildarzadeh 2001) for a detailed review of the method, the Brillouin zone is divided into patches with the average momentum $\boldsymbol{K}_n$ on patch n. The self-energy is then assumed to be constant on each patch and dependent only on the momentum $\boldsymbol{K}_n$. The Dyson equation (1.37) has to be generalized to calculate the lattice Green function on the patches. Accordingly, the generalized bath Green function is defined by using (1.38) for each $\boldsymbol{K}_n$. The generalized impurity problem has to be solved now for n impurities. However, each impurity is not only coupled to a local bath but also to the other impurities by frequency-dependent hopping matrix elements, which makes the solution of the DCA considerably more difficult than the DMFT. The couplings extend across the boundaries of the cluster, i.e., the DCA has periodic boundary conditions.

A different approach is taken in the CDMFT scheme (Lichtenstein and Katsnelson 2000, Kotliar, Savrasov, Pálsson and Biroli 2001). Here, the lattice in real space is divided into clusters of finite size. One of those clusters is then treated explicitly in a mean environment of the surrounding clusters. Since the clusters in CDMFT are free (i.e., non-local correlations are only treated for sites inside a cluster but not between sites in different clusters), the CDMFT has open boundary conditions and the translational symmetry is broken.

Both the DCA and the CDMFT reduce to the DMFT in the limit of single site clusters and both yield causal Green functions. Whether one of the approaches is generally superior is still an open question and likely depends on the system under investigation. Since the CDMFT breaks the translational symmetry which is preserved in DCA, the DCA should provide a better momentum resolution. On the other hand, short range correlations should be better described by CDMFT since the cluster and the lattice Hamiltonian are directly connected in this scheme. Both cluster extensions have been used successfully to study various model systems. For example, within DCA, it was possible to find d-wave superconductivity in the two-dimensional Hubbard model (Lichtenstein and Katsnelson 2000, Maier, Jarrell, Pruschke and Keller 2000). In the context of LDA+DMFT, the cluster DMFT was used to study the metal-insulator transition in Ti_2O_3 (Poteryaev, Lichtenstein and Kotliar 2003).

2. NUMERICAL ASPECTS

2.1 Quantum Monte Carlo simulation

Throughout this work, we use the auxiliary-field quantum Monte Carlo (QMC) algorithm to solve the single-site problem of the DMFT, i.e., Eq. (1.39). It was first formulated by Hirsch and Fye (1986) who studied magnetic impurities in metals. Based on this work, QMC codes for the solution of the DMFT problem were developed by various groups (Jarrell 1992, Rozenberg, Zhang and Kotliar 1992, Georges and Krauth 1992, Ulmke, Janiš and Vollhardt 1995) and extensively applied in DMFT studies. Nowadays, QMC is a well established method which allows us to calculate the impurity Green function $G(\tau)$ (for a given bath Green function $\mathcal{G}(\tau)$ and a set of model parameters) as well as correlation functions of multi-band Hubbard models. It is numerically exact, i.e., by increasing the numerical effort, its error can principally be made arbitrarily small. Since the QMC is formulated in imaginary time τ and in imaginary Matsubara frequencies,[1] dynamical information (e.g., the spectral function $A(\omega)$) on the real axis can only be obtained from analytic continuation of the QMC data (see Sec. 2.3). A further implication of the imaginary-time formulation is the necessity of Fourier transformations in the self-consistency cycle (Fig. 1.5) which connect the Dyson equation (1.37) and Eq. (1.38) that are formulated in Matsubara frequencies with the imaginary-time impurity problem (1.39). The problems connected to the Fourier transformations are discussed in Sec. 2.2. In the QMC algorithm, the interacting electron problem (1.39) is mapped onto a sum of non-interacting problems, see Sec. 2.1.1. This high-dimensional sum is then evaluated by Monte Carlo sampling which is discussed in detail in Sec. 2.1.2. Due to the strong temperature dependence of the computing time which scales with $\frac{1}{T^3}$, the applicability of the QMC code presented here is restricted to $T \gtrsim 300$ K for three-band calculations. Just recently, a novel projective quantum Monte Carlo (PQMC), that is technically closely related to the QMC method presented here, was proposed by Feldbacher, Held and Assaad (2004). This new algorithm can be used as an impurity solver in the framework of DMFT and makes it possible to calculate ground-state properties and, e.g., obtain results for the Mott-Hubbard MIT. Another interesting new development is the continuous-time quantum Monte Carlo method by Rubtsov

[1] QMC calculations are done in imaginary time τ because the real time operator $e^{it\hat{H}}$ is strongly oscillating at large t and makes Monte Carlo sampling inefficient.

et al. (2004) which does not utilize a Hubbard-Stratonovich transformation and allows one to take into account general non-local interactions.

2.1.1 Time discretization and Wicks theorem

In order to solve the impurity problem, the imaginary-time interval $[0, \beta]$ (with inverse temperature $\beta = 1/k_B T$) of the functional integral of Eq. (1.39) is discretized into Λ intervals of size $\Delta\tau = \beta/\Lambda$. With this so-called Trotter discretization, the integral $\int_0^\beta d\tau$ can be replaced by the sum $\sum_{l=1}^{\Lambda} \Delta\tau$. Subsequently, the Trotter-Suzuki formula (Trotter 1959, Suzuki 1976) for operators $\hat{A}$ and $\hat{B}$

$$e^{-\beta(\hat{A}+\hat{B})} = \prod_{l=1}^{\Lambda} e^{-\Delta\tau\hat{A}} e^{-\Delta\tau\hat{B}} + \mathcal{O}(\Delta\tau), \tag{2.1}$$

which is exact in the limit $\Delta\tau \to 0$, can be used to separate the exponential terms $e^{-\Delta\tau\hat{A}}$ and $e^{-\Delta\tau\hat{B}}$, i.e., the kinetic energy and interaction terms in Eq. (1.39). With the discretized single site action $\mathcal{A}$ of Eq. (1.40)[2]

$$\begin{aligned} \mathcal{A}_\Lambda[\psi, \psi^*, \mathcal{G}^{-1}] &= (\Delta\tau)^2 \sum_{m\,\sigma\,l,l'=0}^{\Lambda-1} \psi^{\sigma}_{ml}{}^{*} \mathcal{G}^{\sigma}_{m}{}^{-1}(l\Delta\tau - l'\Delta\tau)\psi^{\sigma}_{ml'} \\ &\quad -\frac{1}{2}\Delta\tau {\sum}'_{m\sigma,m'\sigma'} U^{\sigma\sigma'}_{mm'} \sum_{l=0}^{\Lambda-1} \psi^{\sigma}_{ml}{}^{*}\psi^{\sigma}_{ml}\psi^{\sigma'}_{m'l}{}^{*}\psi^{\sigma'}_{m'l}, \end{aligned} \tag{2.2}$$

and applying (2.1), we obtain to lowest order

$$\begin{aligned} \exp(\mathcal{A}_\Lambda[\psi, \psi^*, \mathcal{G}^{-1}]) &= \prod_{l=1}^{\Lambda-1} \Bigg[\exp\Big((\Delta\tau)^2 \sum_{m\,\sigma\,l'=0}^{\Lambda-1} \psi^{\sigma}_{ml}{}^{*} \mathcal{G}^{\sigma}_{m}{}^{-1}(l\Delta\tau - l'\Delta\tau)\psi^{\sigma}_{ml'} \Big) \\ &\quad \times \exp\Big(-\frac{1}{2}\Delta\tau \sum_{m\sigma,m'\sigma'}{}' U^{\sigma\sigma'}_{mm'}\psi^{\sigma}_{ml}{}^{*}\psi^{\sigma}_{ml}\psi^{\sigma'}_{m'l}{}^{*}\psi^{\sigma'}_{m'l} \Big) \Bigg]. \end{aligned} \tag{2.3}$$

By a shift in chemical potential, the (four-fermion) interaction terms can be brought to a quadratic form (with two-fermion terms). In this form, the $M(2M-1)$ interaction terms of Eq. (2.3) can be decoupled by applying the

[2] Here, the first term of Eq. (1.40) was Fourier-transformed from Matsubara frequencies to imaginary time.

discrete (Hirsch-Fye)-Hubbard-Stratonovich transformation[3] (Hirsch 1983)

$$\exp\left\{\frac{\Delta\tau}{2}U^{\sigma\sigma'}_{mm'}({\psi^{\sigma}_{ml}}^{*}\psi^{\sigma}_{ml}-{\psi^{\sigma'}_{m'l}}^{*}\psi^{\sigma'}_{m'l})^2\right\}=$$
$$\frac{1}{2}\sum_{s^{\sigma\sigma'}_{lmm'}=\pm1}\exp\left\{\Delta\tau\lambda^{\sigma\sigma'}_{lmm'}s^{\sigma\sigma'}_{lmm'}({\psi^{\sigma}_{ml}}^{*}\psi^{\sigma}_{ml}-{\psi^{\sigma'}_{m'l}}^{*}\psi^{\sigma'}_{m'l})\right\}, \tag{2.4}$$

with $\cosh(\lambda^{\sigma\sigma'}_{lmm'})=\exp(\Delta\tau U^{\sigma\sigma'}_{mm'}/2)$ and the number of interacting orbitals M. Here, the electron-electron interaction is replaced by an interaction of the electrons with a classic auxiliary field with $\Lambda M(2M-1)$ components $s^{\sigma\sigma'}_{lmm'}$. For fermions on a lattice, this auxiliary field can be interpreted as an ensemble of Ising spins $\boldsymbol{s}=\{s^{\sigma\sigma'}_{lmm'}\}$ (Hirsch 1983). The Green function between the imaginary times $\tau_1=l_1\Delta\tau$ and $\tau_2=l_2\Delta\tau$, $G(\tau_1-\tau_2)\equiv G_{l_1l_2}$, can now be written as:[4]

$$G^{\tilde{\sigma}}_{\tilde{m}l_1l_2}=\frac{1}{\mathcal{Z}}\frac{1}{2}\sum_{\{\boldsymbol{s}\}}\int\mathcal{D}[\psi]\mathcal{D}[\psi^*]\psi^{\tilde{\sigma}*}_{l_1\tilde{m}}\psi^{\tilde{\sigma}}_{l_2\tilde{m}}\exp\Big(\sum_{m\,\sigma\,ll'}\psi^{\sigma*}_{ml'}M^{\sigma\boldsymbol{s}}_{mll'}\psi^{\sigma}_{ml'}\Big). \tag{2.5}$$

Here, the partition function $\mathcal{Z}$ is

$$\mathcal{Z}=\sum_{\{\boldsymbol{s}\}}\int\mathcal{D}[\psi]\mathcal{D}[\psi^*]\exp\Big(\sum_{m\,\sigma\,ll'}\psi^{\sigma*}_{ml'}M^{\sigma\boldsymbol{s}}_{mll'}\psi^{\sigma}_{ml'}\Big), \tag{2.6}$$

the matrix $\boldsymbol{M}^{\tilde{\sigma}\boldsymbol{s}}_{\tilde{m}}$ has the elements

$$\boldsymbol{M}^{\tilde{\sigma}\boldsymbol{s}}_{\tilde{m}}=\Delta\tau^2[{\boldsymbol{G}^{\sigma}_{m}}^{-1}+\Sigma^{\sigma}_{m}]e^{-\tilde{\boldsymbol{\lambda}}^{\sigma\boldsymbol{s}}_{m}}+\boldsymbol{1}-e^{-\tilde{\boldsymbol{\lambda}}^{\sigma\boldsymbol{s}}_{m}} \tag{2.7}$$

and the matrix $\tilde{\boldsymbol{\lambda}}^{\sigma\boldsymbol{s}}_{m}$ is defined as

$$\tilde{\lambda}^{\sigma\boldsymbol{s}}_{mll'}=-\delta_{ll'}\sum_{m'\sigma'}\lambda^{\sigma\sigma'}_{mm'}\tilde{\sigma}^{\sigma\sigma'}_{mm'}s^{\sigma\sigma'}_{lmm'}. \tag{2.8}$$

If $(m\sigma)$ and $(m'\sigma')$ are exchanged, $\tilde{\sigma}^{\sigma\sigma'}_{mm'}\equiv2\Theta(\sigma'-\sigma+\delta_{\sigma\sigma'}[m'-m]-1)$ changes sign. More details on the derivation of Eq. (2.7) for the matrix $\boldsymbol{M}$ can be found in Georges et al. (1996) and Jarrell (1997).

Since the fermion operators now only enter quadratically, i.e., the system is non-interacting for a given configuration $\boldsymbol{s}$, the functional integral can be solved by applying Wick's theorem (see, e.g., Negele and Orland (1987)) which reduces the quantum mechanical problem to a matrix problem

$$G^{\tilde{\sigma}}_{\tilde{m}l_1l_2}=\frac{1}{\mathcal{Z}}\frac{1}{2}\sum_{\{\boldsymbol{s}\}}\left[(M^{\tilde{\sigma}\boldsymbol{s}}_{\tilde{m}})^{-1}\right]_{l_1l_2}\prod_{m\sigma}\det\boldsymbol{M}^{\sigma\boldsymbol{s}}_{m}. \tag{2.9}$$

[3] This transformation can be applied to the Coulomb repulsion and the z-component of the Hund's rule coupling, but it fails for the full Hund term.

[4] Here, the sum $\sum_{\{\boldsymbol{s}\}}\equiv\sum_l\sum'_{m'\sigma',m''\sigma''}\sum_{s^{\sigma''\sigma'}_{lm''m'}=\pm1}$ extends over all Ising spins of the auxiliary field.

This sum consists of $2^{\Lambda M(2M-1)}$ addends, the computation of each addend from Eq. (2.7) is a operation of order $\mathcal{O}(\Lambda^3)$.[5] It is clear that such a exponential sum can only be calculated exactly for very small Λ, i.e., at extremely high temperatures or for very coarse discretizations $\Delta\tau = \beta/\Lambda$.[6] With the Monte Carlo method, which will be described in the next section, it is possible to drastically reduce the numerical effort and do computations for larger Λ.

2.1.2 Monte Carlo importance sampling

Monte Carlo (MC) simulations are a widely used tool in computational physics and related fields for solving various kinds of problems stochastically, i.e., by generating random configurations and obtaining the mean behavior of the system from the distribution of those configurations. When the random configurations are spread uniformly, the distribution becomes Gaussian in the limit of infinite number of configurations according to the central limit theorem and the MC yields the exact result (i.e., the error goes to zero). Monte Carlo methods are especially suitable to estimate large sums (or high-dimensional integrals), because even for a relatively small number of terms of the sum (or a small number of points where the integrand is evaluated), the MC approach can yield a good estimate for the full sum (or integral). We therefore employ such a MC algorithm to calculate the high-dimensional sum of Eq. (2.9). To get faster convergence to a Gaussian distribution, importance sampling (a variance reduction technique) is used. In this method, the function $f(x_i)$ (on the set of MC configurations $x_1, ..., x_N$) is split up into a factor $p(x_i)$ with

$$\sum_{i=1}^{N} p(x_i) = c \qquad \text{and} \qquad p(x_i) \geq 0, \tag{2.10}$$

which can be viewed as a (unnormalized) probability distribution, and a remaining observable $o(x_i)$. If the MC configurations x_i can be generated so that they obey the probability distribution[7] $P(x) = p(x_i)/c$ and if the normalization c can be calculated, the sum over $f(x_i)$ can be written as

$$\sum_{i=1}^{N} f(x_i) = \frac{c}{N} \sum_{i=1}^{N} o(x_i) \equiv < o >_P \tag{2.11}$$

[5] The cost of this operation can be reduced to $\mathcal{O}(\Lambda^2)$, if the terms are ordered such that successive configurations $\{s\}$ and $\{s'\}$ are differing by only one spin-flip (Blankenbecler, Scalapino and Sugar 1981).

[6] Typical calculations in this work were done with $\Lambda = 40 - 166$; in a three-band model, exact summations are only possible for $\Lambda \lesssim 2$.

[7] In statistical physics, a good choice for this function $P(x)$ is often the Boltzmann distribution $P(x) = \frac{1}{Z}\exp(-\beta E(x))$.

with an error $\Delta_o = c\sqrt{N}\sqrt{< o^2 >_P - < o >_P^2}$. Hence, configurations x_i in a relevant region of the phase space (i.e., where the probability $P(x)$ is large) are contributing to a greater extent to the sum than configurations in a region with low probability. Since the normalization c is generally not known, the probability distribution $P(x)$ is realized in a stochastic Markov process. In this process, a chain of configurations x_i is built from a starting configuration x_1. For each configuration x_i, a transition is only possible to a certain subset of configurations $x_{i'}$. If the process is ergodic (i.e., all possible configurations can be reached from every starting configuration) and the transitions fulfill the detailed balance

$$p(x_i)\mathcal{P}(x_i \to x_{i'}) = p(x_{i'})\mathcal{P}(x_{i'} \to x_i), \tag{2.12}$$

the configurations realized by the Markov process are distributed like $P(x)$ in the limit of infinite chain length. In this work, the Metropolis transition rule (Metropolis, Rosenbluth, Rosenbluth, Teller and Teller 1953) for a transition from configuration x to y which fulfills the detailed balance (2.12) is used:

$$\mathcal{P}(x \to y) = \min(1, P(y)/P(x)). \tag{2.13}$$

As the normalization c is not known, the Markov process cannot yield absolute values for observables, only ratios of different observables can be computed. Furthermore, if the (arbitrary) start configuration and the following configurations have a negligible probability, they will be overrepresented in the sum and therefore have to be excluded from the average. Those first MC updates will later be called warm-up sweeps to distinguish them from the measurement sweeps from which the average is obtained. A further consequence of the use of the Markov process is that subsequent configurations are correlated. Hence, a larger number of configurations has to be computed to reach an intended statistical accuracy. For a given number of configurations, the error therefore is larger than for a statistically independent ensemble. This is expressed by the autocorrelation time $\kappa_0 \geq 0$ which leads to an error $\Delta_o = c\sqrt{\kappa_0/N}\sqrt{< o^2 >_P - < o >_P^2}$, for example, see Blümer (2002).

For the high-dimensional sum of Eq. (2.9) that is to be calculated, the normalized probability distribution (called $P(x)$ before) can be identified with

$$P(\boldsymbol{s}) = \frac{1}{\mathcal{Z}} \prod_{m\sigma} \det \boldsymbol{M}_m^{\sigma \boldsymbol{s}}, \tag{2.14}$$

the observable then is

$$O(\boldsymbol{s})_{\tilde{m}l_1l_2}^{\tilde{\sigma}} = \left[(M_{\tilde{m}}^{\tilde{\sigma}\boldsymbol{s}})^{-1} \right]_{l_1l_2}. \tag{2.15}$$

The partition function $\mathcal{Z}$ that enters the product in (2.14) corresponds to the normalization factor c. Due to the importance sampling, it is therefore not possible to calculate the partition function in QMC which restricts its use in studies

Choose random auxiliary field configuration $\boldsymbol{s} = \{s_{lmm'}^{\sigma\sigma'}\}$

Calculate the current Green function $\boldsymbol{G}_{\rm cur}$ from Eq. (2.15)

$$(G_{\rm cur})^{\tilde{\sigma}}_{\tilde{m}l_1l_2} = \left[(M^{\tilde{\sigma}\boldsymbol{s}}_{\tilde{m}})^{-1}\right]_{l_1l_2}$$

with $\boldsymbol{M}$ from Eq. (2.7) and the input $\mathcal{G}^\sigma_m(\omega_\nu)^{-1} = G^\sigma_m(\omega_\nu)^{-1} + \Sigma^\sigma_m(\omega_\nu)$.

Do NWU times (warm up sweeps)

MC-sweep $(\boldsymbol{G}_{\rm cur}, \boldsymbol{s})$

Do NMC times (measurement sweeps)

MC-sweep $(\boldsymbol{G}_{\rm cur}, \boldsymbol{s})$

$\boldsymbol{G} = \boldsymbol{G} + \boldsymbol{G}_{\rm cur}/\mathrm{NMC}$

Figure 2.1: Flow diagram of the QMC algorithm to calculate the Green function matrix $\boldsymbol{G}$ using the procedure *MC-sweep* of Fig. 2.2. Figure from Held et al. (2003).

of phase transitions (Blümer 2002). If Eq. (2.14) includes negative determinants, the absolute value of the determinant has to be taken, its sign is then considered in the observable of Eq. (2.15). If the number of such negative contributions to the sum (2.11) is large, the sum and thus the average $< o >_P$ will be small, whereas the error Δ_o is large. This is the so-called minus-sign problem that seriously restricts the applicability of Monte Carlo simulations to fermion systems (e.g., for a full Hund's rule coupling term as described in Sec. 1.2.1).

A flow diagram of the complete Monte Carlo algorithm used in this work is shown in Fig. 2.1.

To minimize the computing cost, the configuration updates ($x \to y$) are done by flipping one spin of the auxiliary field $s^{\sigma\sigma'}_{lmm'}$ and accept it only with the probability $\mathcal{P}(x \to y)$ (Eq. 2.13). In one MC-sweep, such a spin flip is attempted for each component of the auxiliary field, see the flow diagram of Fig. 2.2. This procedure, which is discussed in detail by Held (1999), reduces the computing cost for the update of $\boldsymbol{G}_{\rm cur}$ (i.e., the matrix $\boldsymbol{M}$ of Eq. (2.7)), to order $\mathcal{O}(\Lambda^2)$ (Hirsch and Fye 1986, Georges et al. 1996). With the number of components of the auxiliary field $\Lambda M(2M-1)$, the total numerical cost of the algorithm to leading order of Λ is

$$2\mathcal{A}M(2M-1)\Lambda^3 \times \text{number of MC-sweeps,} \tag{2.16}$$

where $\mathcal{A}$ is the acceptance rate for a single spin-flip.

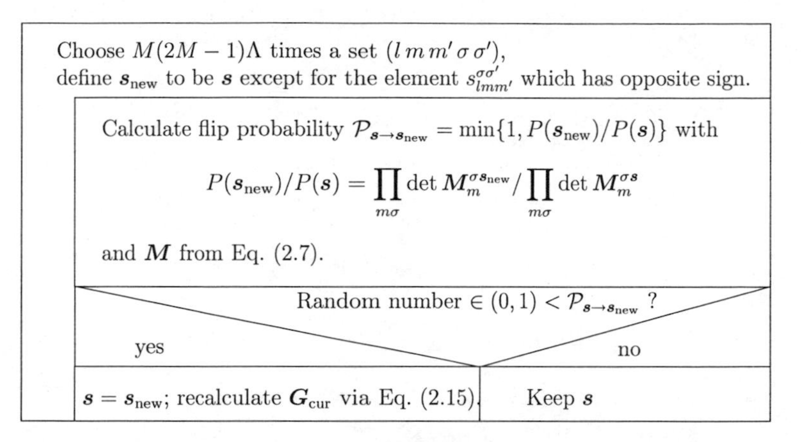

Figure 2.2: Procedure *MC-sweep* using the Metropolis rule (Metropolis et al. 1953) to change the sign of $s^{\sigma\sigma'}_{lmm'}$. Figure from Held et al. (2003).

2.2 The problem of Fourier transformations and its solutions

In the self-consistency cycle of the DMFT presented in Fig. 1.5, the Dyson equation (1.37) is formulated in fermionic Matsubara frequencies ω_ν whereas the bath Green function $\mathcal{G}$ of the impurity problem (1.39) is expressed as a function of imaginary time τ. Therefore, it is necessary to perform two Fourier transformations per iteration of the DMFT cycle, one from frequency to imaginary time after the solution of the Dyson equation and the calculation of the bath Green function $\mathcal{G}$, and one from imaginary time to frequency after solving the impurity problem with QMC:[8]

$$G(\tau) = \frac{1}{\beta} \sum_{\nu=-\infty}^{\infty} e^{-i\omega_\nu \tau} G(i\omega_\nu), \tag{2.17}$$

$$G(i\omega_\nu) = \int_0^\beta d\tau e^{i\omega_\nu \tau} G(\tau). \tag{2.18}$$

Since the Green function in DMFT is antiperiodic for translations β, $G(-\tau) = -G(\beta-\tau)$ and its representation in Matsubara frequencies fulfills $G(i\omega_\nu) =$

[8] In the QMC code, the transformation (2.17) is performed for the bath Green function $\mathcal{G}$, but for simplification, we present both Fourier transformations for the lattice Green function G.

$G^*(-i\omega_\nu)$.[9] Furthermore, $G(i\omega_\nu)$ decays as $1/\omega_\nu$ for $|\nu| \to \infty$. In Eq. (2.17), the number of terms of the sum is infinite, hence $G(\tau)$ can have a discontinuity at $\tau = 0$.

In order to numerically perform those Fourier transformations in the QMC code, the sum over Matsubara frequencies in (2.17) has to be truncated and the integral in (2.18) has to be discretized. A straightforward Fourier transform with symmetric discretization (i.e., with equal number Λ of Matsubara frequencies ω_ν and time slices $\tau_l = l\Delta\tau$) would have the following form (Blümer 2002):

$$\tilde{G}(\tau_l) = \frac{1}{\beta} \sum_{\nu=-\Lambda/2}^{\Lambda/2} e^{-i\omega_\nu \tau_l} G(i\omega_\nu), \tag{2.19}$$

$$\tilde{G}(i\omega_\nu) = \Delta\tau \Big(\frac{G(0) - G(\beta)}{2} + \sum_{l=1}^{\Lambda-1} e^{i\omega_\nu \tau_l} G(\tau_l) \Big). \tag{2.20}$$

Since the numerical cost of the QMC scales with Λ^3, the number of time slices is restricted to about 400 for one-band QMC calculations and about 160 for three-band calculations (most of the calculations presented in this work were done for three bands with 40, 66 and 166 time slices). $\tilde{G}(i\omega_\nu)$ shows an oscillation with periodicity $2\pi i\Lambda/\beta$ instead of a $1/i\omega_\nu$ decay for high frequencies which leads to a large error of $G(i\omega_\nu)$ for frequencies around or above the Nyquist frequency $\pi\Lambda/\beta$, especially for a small number of time slices. Another problem is that the discontinuity at $\tau = 0$ of the true $G(\tau)$ cannot be reproduced in the truncated Matsubara sum of $\tilde{G}(\tau)$ and that $\tilde{G}(\tau)$ also shows oscillations between the grid points τ_l. In conclusion, such a straightforward Fourier transform does not work for the coarse grid that is necessary to be able to perform QMC calculations. There are different approaches to cure the deficiencies of this naive scheme.

Ulmke (1995) proposed a combination of the naive Fourier transform described above with a smoothing transformation that enforces the correct analytical behavior and is equal to one in the limit $\Delta\tau \to 0$. Instead of the symmetric Fourier transformation (2.20), an asymmetric version is used, $\tilde{G}(i\omega_\nu) = \Delta\tau \sum_{l=0}^{\Lambda 0} e^{i\omega_\nu \tau_l} G(\tau_l)$ which differs from the symmetric form by a constant factor of $\Delta\tau/2$. After the Fourier transformation, the smoothing transformation

$$G(i\omega_\nu) = \frac{\Delta\tau}{\ln(1 + \Delta\tau/\tilde{G}(i\omega_\nu))} \tag{2.21}$$

is applied. With this correction, the Green function $G(i\omega_\nu)$ has (approximately) the correct behavior at large frequencies and decays as $1/\omega_\nu$ for $\omega \to \pm\omega_{max}$. For symmetric density of states and half-filled bands, it furthermore obeys

[9] In the half-filled case, $G(i\omega_\nu)$ is purely imaginary and $G(\tau) = G(\beta - \tau)$.

Re $G(i\omega_\nu) = 0$. With this corrected Green function, the Dyson equation is solved. Subsequently, the inverse smoothing transformation

$$G(i\omega_\nu) = \frac{\Delta\tau}{\exp(\Delta\tau/\tilde{G}(i\omega_\nu) - 1)} \tag{2.22}$$

and after that the inverse Fourier transformation (2.19) are applied, yielding a imaginary-time Green function $G(\tau)$ which is correct for $\tau \to 0$ and fulfills $G(\beta) = 1 - G(0)$. Since the correction of $G(i\omega_\nu)$ due to the smoothing transformation is not frequency dependent, all values of the Green function that are comparable to or smaller than $\Delta\tau$ are significantly changed. Thus, the smoothing does not only correct $G(i\omega_\nu)$ for large frequencies $\omega \to \pm\omega_{max}$ as desired but also erroneously changes the small Green function values at small frequencies encountered for an insulating system. In consequence, metallic solutions are artificially favored and the insulating solution partly suppressed. Furthermore, the $\Delta\tau$ dependency of results obtained with Ulmke's smoothing transformation is distinctively stronger than for other Fourier transformation schemes. Blümer (2002) studied the metal-insulator transition and the coexistence of the insulating and metallic solution in detail for the one-band Hubbard model. Furthermore, he proposed an improved smoothing transformation that preserves the general idea of Ulmke's approach but corrects its main deficiencies. To this end, the $\Delta\tau$ factors in (2.21) are replaced by $\alpha(\omega_\nu)\Delta\tau$ with

$$\alpha(\omega_\nu) = 1 - (\omega_\nu\Delta\tau/\pi - 1)^8. \tag{2.23}$$

With the additional factor $\alpha(\omega_\nu)$, which goes to zero for $\omega_\nu \to 0$ and rapidly approaches the value 1 for frequencies of the order of the Nyquist frequency $\pi/\Delta\tau$, the smoothing transformation still works as desired for large frequencies while it does not negatively affect the low frequency behavior. For small ω_ν, the error is of order $\mathcal{O}(\Delta\tau^2)$ instead of $\mathcal{O}(\Delta\tau)$ as in the Ulmke smoothing which is a dramatic improvement. In the calculations of this work, the original Ulmke smoothing transformation was used. The values of the Green function for low frequencies were found to be larger than $\Delta\tau$ in all cases, thus the smoothing artefact of the Ulmke transformation did not lead to problems. This was verified by performing again selected calculations for V_2O_3 (near the MIT and in the insulating phase) with the Blümer smoothing transformation and finding practically identical results (with only minimal differences which can be attributed to the statistical nature of the QMC).

In the naive scheme as well as in Ulmke's scheme (and in the improved version by Blümer), the number of Matsubara frequencies and the number of time slices is equal. However, only in the impurity problem (1.39), the number of time slices is restricted to a low number due to the QMC (e.g., $\Lambda \lesssim 160$ for three-band calculations). The Dyson equation (1.37) is computationally uncritical and can be solved for a much larger number of Matsubara frequencies ($\nu \approx 1000 - 10000$). Thus, the Green function $G(i\omega_\nu)$ can be calculated in more detail and for frequencies above the Nyquist frequency. In order to obtain $G(i\omega_\nu)$, one can exploit that $G(\tau)$ is known to be a smooth function and that $G(\tau)$ and all even derivatives are positive definite and hence maximal at $\tau = 0$ and $\tau = \beta$ by interpolating $G(\tau_l)$ from QMC ($l = 0, ..., \Lambda$). The interpolation is often done with splines, i.e., with piecewise defined cubic polynomials, but other basis functions are also used. The piecewise defined functions resulting from the interpolation can then either be used for a direct analytical Fourier transformation or they can be used to generate $G(\tau)$ on a larger (but still finite) grid and do a transformation for each grid point (oversampling). The first possibility is realized in the QMC code by Georges et al. (1996) for splines (with typically $\nu \approx 8000$) and by McMahan, Held and Scalettar (2003) for exponential functions. Jarrell, Akhlaghpour and Pruschke (1993) implemented the Fourier transformation with oversampling and typically use 800 Matsubara frequencies. Instead of cubic splines, which were found to generate artificial high-frequency features in the spectra, Jarrell uses Akima splines which reduces those artifacts. For such splines the second derivative vanishes at the end points, which leads to a bad fit to the real $G(\tau)$ where the second derivatives are maximal at the end points and hence strong oscillations of the interpolated function. In Jarrell's code, this problem is solved by doing the Fourier transformation not for the Green function $G(\tau_l)$ itself, but for the difference between the QMC Green function and a reference Green function obtained from a iterative perturbation theory (IPT) calculation.[10] An oversampling scheme similar to Jarrells' was proposed by Knecht (2002) and Blümer (2002). Instead of the IPT result, it uses a reference function based on an exact high-frequency expansion for the self-energy by Potthoff, Wegner and Nolting (1997), which makes it more stable as it does not rely on IPT. Since the number of Matsubara frequencies in all the interpolation schemes presented here is much larger than the number of time slices, the Fourier transformation back to imaginary time is unproblematic. A comparison of different splining and smoothing Fourier transformation schemes with a detailed analysis of the advantages and drawbacks of the respective methods can be found in Blümer (2002).

[10] Furthermore, for high frequencies, the QMC estimate is supplemented by the IPT result in the Jarrel code.

2.3 Maximum entropy method

From the local retarded Green function $G_R(t)$ or rather from its spectral function $A(\omega)$ which is defined as

$$A(\omega) = -\frac{1}{\pi}\text{Im } G_R(\omega) \tag{2.24}$$

with the Fourier transformation $G_R(\omega) = \int_0^\infty \frac{dt}{2\pi} e^{i\omega t} G_R(t)$, all single particle properties of the system can be obtained. Of special interest are experimentally accessible quantities. Under certain assumptions (see, e.g., Blümer (2002)), the spectral function $A(\omega)$ itself multiplied with the Fermi function or the inverse Fermi function can be directly compared with experimental photo-emission spectroscopy (PES) (Cardona and Ley 1978) and x-ray absorption spectroscopy (XAS), respectively. The calculation of theoretical PES and XAS spectra are discussed in detail in appendix A.

The quantum Monte Carlo algorithm presented in Sec. 2.1 is formulated in imaginary time. Thus, the Green function G cannot be calculated directly on the real axis, only imaginary-time data $G(\tau)$ or Matsubara frequency data $G(i\omega_\nu)$, respectively, can be obtained. In order to extract dynamical information on the real axis from this data, an analytic continuation has to be performed. The spectral function on the real axis is linked to the imaginary Green function via[11]

$$G(\tau) = \int_{-\infty}^{\infty} d\omega \frac{e^{-\tau\omega}}{1+e^{-\beta\omega}} A(\omega) \quad , \quad G(i\omega_\nu) = \int_{-\infty}^{\infty} d\omega' \frac{A(\omega)}{i\omega_\nu - \omega'}. \tag{2.25}$$

In principle, this spectral representation of $G(\tau)$ can be inverted to obtain $A(\omega)$, however, this inversion is ill-conditioned. The fermion kernel $K(\tau,\omega) = \frac{e^{-\tau\omega}}{1+e^{-\beta\omega}}$ of Eq. (2.25) is exponentially small at large positive and negative frequencies. Therefore changes in $A(\omega)$ in this frequency region have only a minimal effect on the Green function. On the other hand, since $G(\tau)$ can only be measured on the grid points $\tau_l = l\Delta\tau$ (with $l = 0, ..., \Lambda$) and with a statistical error, even small changes in the Green function or a coarse $\Delta\tau$ grid will strongly affect the spectral function at large ω. Only for small frequencies, $A(\omega)$ can be calculated accurately.[12] Hence, for a given $G(\tau)$ from QMC, there exists an infinite number of spectral functions that fulfill Eq. (2.25). The spectral density has some general features that are important in the context of the maximum entropy. It is positive $A(\omega) \geq 0$ and bounded $\int_{-\infty}^{\infty} d\omega A(\omega) < \infty$ and can thus be redefined to a normalized distribution. This allows us to interpret the spectral function as a probability function.

[11] In the following, we will concentrate on the imaginary-time Green function $G(\tau)$.

[12] According to the Nyquist theorem (also called Nyquist-Shannon sampling theorem), no information can be obtained from QMC for frequencies above the Nyquist frequency $\omega > \pi\Lambda/\beta$.

Various methods have been employed to address the analytic continuation problem for QMC data with only limited success. The Padé approximation, that can be used to find an explicit analytic form of the Green function (from which the analytic continuation is trivial), works well only for very precise data, but not for the incomplete and noisy QMC results. Least-squares fits are inherently unstable and produce spurious (high-frequency) features in the spectra. The maximum entropy method (MEM) has proven to be the method of choice for the analytic continuation of imaginary-time QMC results. Its basic idea is to interpret the spectral density as a probability function and to find the most probable density of states that complies with the data. An interesting new alternative to the MEM is the stochastic analytic continuation which was proposed by Sandvik (1998). In this promising method, the analytic continuation problem is mapped to a system of interacting classical fields and a thermal average over the field configurations is performed using Monte Carlo sampling. Its main advantage over the MEM is that it can better resolve sharp features and flat regions in the spectrum. Recently, it was shown by Beach (2004) that the MEM is a special limit of this new approach.

MEM has a long history in statistical data analysis, especially for image reconstruction in various fields ranging from medical tomography to radio astronomy and x-ray imaging (for details and further references, see Skilling and Bryan (1984)). Its applicability to the analytic continuation of QMC data was first studied by Silver, Sivia and Gubernatis (1990), a general MEM algorithm for the analytic continuation of QMC data was developed by Gubernatis, Jarrell, Silver and Sivia (1991), based on the work of Bryan (1990).

In the QMC simulations, we measure the Green function $\tilde{G}(\tau)$ that deviates from the exact Green function $G(\tau) = \int d\omega K(\tau,\omega)A(\omega)$ due to statistical and systematic errors. One can now define a functional

$$\chi^2[A] = \int_0^\beta \frac{d\tau}{\sigma^2(\tau)} |G(\tau) - \tilde{G}(\tau)|^2, \tag{2.26}$$

which gives a measure of how well the Green function generated from $A(\omega)$ matches the function $\tilde{G}(\tau)$ from QMC. $\sigma(\tau)$ is an estimate for the total measurement error in $\tilde{G}(\tau)$. If one disregards systematic errors and the oversampling[13] in the QMC, assumes that there is no autocorrelation in the QMC data and further implies a Gaussian distribution of the statistical errors, the exact Green function can be approximated by the average of the measured Green function. For

[13] According to terminology of Bryan (1990), data is oversampled if there is a correlation between measurements of different data points (in our case for different τ). Thus, those measurements are not statistically independent which has a negative effect on the error statistics.

each time slice l, the average is $\bar{G}_l = \frac{1}{N}\sum_{i=1}^{N} G_l^i$ (with the QMC Green function $G_l = G(l\Delta\tau)$, $0 \leq l < \Lambda$) and the functional χ^2 can be written as

$$\chi^2 = \sum_{l=0}^{\Lambda-1} \frac{1}{\sigma_l^2}(\bar{G}_l - G_l)^2. \tag{2.27}$$

Here, σ_l is the usual standard deviation, $\sigma_l^2 \approx \frac{1}{(N-1)}\sum_{i=1}^{N}(\bar{G}_l - G_l^i)^2$ which can be estimated from the data. We can now define the likelihood function

$$P(\tilde{G}|A) \propto e^{-\frac{1}{2}\chi^2} \tag{2.28}$$

which is the conditional probability of finding $\tilde{G}$ for a given spectral function A. It is maximized when χ^2 is minimized. Since we want to obtain the most probable spectral function for the given (measured) $\tilde{G}$, the converse conditional probability $P(A|\tilde{G})$ is needed. This can be gained from $P(\tilde{G}|A)$ with Bayes' theorem

$$P(A|\tilde{G})P(\tilde{G}) = P(\tilde{G}|A)P(A). \tag{2.29}$$

Here, $P(A|\tilde{G})$ is called the posterior probability, $P(A)$ the prior probability and $P(\tilde{G})$ the evidence. In the determination of A, the QMC data $\tilde{G}$ is constant, thus $P(\tilde{G})$ can be neglected. If one makes the assumption that the prior probability $P(A)$ is not important and can be ignored, $P(A|\tilde{G}) \propto P(\tilde{G}|A)$, this so-called method of maximum likelihood corresponds to the least squares fitting procedure. However, in many cases this leads to noisy and non-unique results. In order to incorporate a regularization in the MEM, the spectral function is interpreted as a probability function and an entropic ansatz is made for the prior probability. The spectral function $A(\omega)$ is chosen to maximize the entropy

$$S[A, D] = \int d\omega \left(A(\omega) - D(\omega) - A(\omega)\ln(A(\omega)/D(\omega))\right), \tag{2.30}$$

which is a measure of the information content of $A(\omega)$. Here, $D(\omega)$ is the so-called default model. It is normally a smooth (positive semidefinite) function that can be chosen to include prior knowledge of spectral features of $A(\omega)$. If $A(\omega)$ and $D(\omega)$ have the same normalization, Eq. (2.30) reduces to $S[A, D] = -\int d\omega \left(A(\omega)\ln(A(\omega)/D(\omega))\right)$. In the absence of data, the posterior probability is given by the prior probability, $P(A|\tilde{G}) \propto P(A)$. In this case, the entropy S attains its maximum value of zero and the spectral function $A(\omega)$ is equal to the default model $D(\omega)$. Hence, the spectral function that maximizes the entropy S also maximizes the posterior probability, $P(A) \propto e^{\alpha S[A,D]}$ with Lagrange parameter α. The posterior probability can thus be written as

$$P(A|\tilde{G}, D, \alpha) = e^{\alpha S[A,D] - \frac{1}{2}\chi^2[\tilde{G},A]}. \tag{2.31}$$

Here, the Lagrange parameter α is used to balance between high entropy and close conformance with the data. For $\alpha \to \infty$, the entropy is maximized and

$A(\omega) = D(\omega)$, whereas $\alpha \to 0$ yields the noisy, unregularized spectrum from the least squares fitting procedure. The minimization of the exponent in (2.31) can be treated as a numerical optimization problem, it is usually achieved using the Newton-Raphson algorithm. Regarding the choice of α, different procedures have been proposed. In historic MEM, the Lagrange parameter is adjusted so that $\chi^2 = \Lambda$, which is the expectation value of χ^2 for Gaussian error distribution. Another possibility, called classic MEM, is to maximize $P(A|\tilde{G}, D, \alpha)$ with respect to A and simultaneously maximize the probability $P(\alpha|\tilde{G}, A, D)$ with respect to α. In a similar scheme (Bryan's approach), $P(A|\tilde{G}, D, \alpha)$ is maximized for each value of α, yielding a set of results A_α. The solution is then defined as $A(\omega) \equiv \int d\alpha A_\alpha P(\alpha|\tilde{G}, A, D)$. This method reproduces the default model in the absence of data, but it does not necessarily lead to a solution that maximizes $P(A|\tilde{G}, D, \alpha)$.

In the calculations presented in this work, we utilized a MEM code by Sandvik and Scalapino (1995) which is based on the algorithm outlined above and obtains α in classic MEM. We obtained spectra mostly with a flat default model ($D(\omega) =$ const.), but we also compared spectra for a flat and a Gaussian default model and found only minor differences, which is an evidence for the good quality of the data.

A technically more refined MEM code that drops some of the simplifying assumptions made above was proposed by Jarrell and Gubernatis (1996). This algorithm takes into account the full covariance matrix, which describes the correlations between measurements and reduces to its diagonal elements only in the uncorrelated case. It is based on Bryan's approach and can give an estimate for error bars. A comparison of the MEM codes by Sandvik and Jarrell and a detailed analysis of the potential problems connected with the analytic continuation of imaginary-time QMC data by MEM can be found in Blümer (2002).

2.4 Computational considerations for QMC simulations

In this section, some more technical points on the realization of QMC simulations will be discussed. Due to the computational effort of the QMC code proportional to $2M(2M-1)\Lambda^3$, calculations for realistic temperatures $T \approx 300$ K for multi-band systems (3 or 5 bands for d-systems, 7 bands for f-systems) require huge computing resources. This is especially true for realistic LDA+DMFT calculations where the number of Monte Carlo sweeps has to be strongly increased to reach convergence. Near a metal-insulator transition, a critical slowing down of the DMFT convergence process can arise which further increases the computation time.

In the course of this work, a parallelization of the multi-band QMC code

(Sec. 2.4.1) has been implemented (analogous to the parallelization of the one-band code by Blümer (2002)) which makes it possible to use a wider range of computers and also decreases the real time necessary for the calculations. By measuring the performance of the code for different system sizes on various computer architectures as shown in Sec. 2.4.2, the optimal system size for each architecture could be found in order to use the available computing resources with maximal gain. The original Fortran77 code by Held (2000) was ported to Fortran90, which makes it easier to extend and maintain and also allows the use of more advanced programming constructs. Further optimizations were implemented by partially unrolling time critical loops or using optimized library routines for certain matrix operations. The parallelized multi-band QMC code was tested and used for computations on a wide variety of computer systems, among others Cray T90, Cray T3E, Fujitsu VPP, Hitachi SR8000-F1, IBM Regatta p690, Sunfire 6800, IBM RS/6000 SP and various (Linux) PC clusters with AMD and Intel processors.

The calculations presented in this work were done with three bands for temperatures between 2300 K and 300 K ($\beta = 5, ..., 42$) with the majority of calculations for 1160 K ($\beta = 10$) and 300 K ($\beta = 42$). The number of time slices was between 40 and 166 with a typical time discretization $\Delta\tau = 0.25$. Up to two million Monte Carlo sweeps were used per iteration for the calculation near the metal-insulator transition of V_2O_3 at high temperatures ($\beta = 10$), up to 200000 sweeps for room temperature ($\beta = 42$). Typically, 10-20 iterations were needed to reach a well-converged solution. Subsequently, up to 10^7 measurement sweeps were performed.

2.4.1 Parallelization of the QMC code

As described above, realistic multi-band LDA+DMFT calculations at experimentally relevant temperatures can only be done on very fast computers. Even on those machines, a parallelization of the code is advantageous to bring the time necessary for the computations to a reasonable level or to go to lower temperatures with the same level of accuracy as in the single processor calculations. The DMFT(QMC) algorithm is well suited for a coarse-grained parallelization, since over 99% of the computing time is spent on the solution of the impurity problem (1.39). Hence, this impurity problem is distributed to all parallel processes and solved independently of each other.[14] In a master process, the Green functions of the parallel processes are collected and averaged. The self-consistency loop which includes the Dyson equation is then evaluated by this master process with the averaged Green function. This is depicted in the parallel version of the DMFT flow diagram in Fig. 2.3. The distribution of the data to the parallel processes and

[14] The random number generator must produce independent random numbers on different processors to ensure independence of the parallel processes.

Choose an initial self-energy Σ

Calculate G from Σ via the k-integrated Dyson Eq. (1.37):

$$G^{\sigma}_{qlm,q'l'm'}(\omega) = \frac{1}{V_B}\int d^3k \left(\left[\,\omega\mathbb{1} + \mu\mathbb{1} - H^0_{\rm LDA}(\boldsymbol{k}) - \Sigma^{\sigma}(\omega)\right]^{-1}\right)_{qlm,q'l'm'}$$

$\mathcal{G} = (G^{-1} + \Sigma)^{-1}$

Distribute data to parallel processes (MPI)

Calculate G from $\mathcal{G}$ via the DMFT single-site problem Eq. (1.39)

$$G^{\sigma}_{\nu m} = -\frac{1}{\mathcal{Z}}\int \mathcal{D}[\psi]\mathcal{D}[\psi^*]\psi^{\sigma}_{\nu m}\psi^{\sigma *}_{\nu m}e^{\mathcal{A}[\psi,\psi^*,\mathcal{G}^{-1}]} \qquad \text{(PARALLEL)}$$

Collect data from parallel processes, calculate averaged G (MPI)

$\Sigma_{\rm new} = \mathcal{G}^{-1} - G^{-1}$

Iterate with $\Sigma = \Sigma_{\rm new}$ until convergence, i.e., $||\Sigma - \Sigma_{\rm new}|| < \epsilon$

Figure 2.3: Parallel version of the flow diagram of the DMFT self-consistency cycle of Fig. 1.5. (PARALLEL): parallel part; (MPI): communication via MPI.

the collection and averaging of the Green function resulting from the solution of the impurity problem are done with the Message Passing Interface (MPI). Since the beginning of its development in 1993, MPI has become the interface of choice for communication-based parallel programs. It is easy to use and available on most supercomputers as well as on cluster systems, which makes the code more portable and allows us to run it on a wide variety of systems. Since the warm-up sweeps for the auxiliary Ising field in the QMC have to be performed for each parallel process, the number of processors that can be used efficiently is limited to about 16 to 32. To utilize more processors efficiently, a fine-grain parallelization (e.g., on the level of the single spin-flips) would be necessary. The main advantage of the coarse-grain parallelization described here is the practically negligible communication overhead which even allows for a efficient utilization of heterogeneous computer clusters with a slow interconnecting network (e.g., ethernet). An important consideration for the parallel version of the program is the choice of a random number generator that produces independent streams of random numbers for each parallel process. In this work, the Scalable Parallel Random Number Generator (SPRNG) library (version 1.0) (Ceperley, Mascagni, Mitas, Saied and Srinivasan 1998) was used.

2.4.2 Performance measurements

For extensive calculations that require large amounts of supercomputer time like the QMC simulations presented in this work, it is mandatory to measure the performance of the code for different system sizes and for different numbers of processors to get the optimal performance and use the available computing time efficiently. In Fig. 2.4, the calculation time on a single processor/node[15] for a three-band QMC simulation is shown for the typical system sizes (i.e., temperatures) used in this work. Fig. 2.5 shows the same data for small system sizes ($\Lambda = 40, 66$) with a different scaling. The computer architectures compared

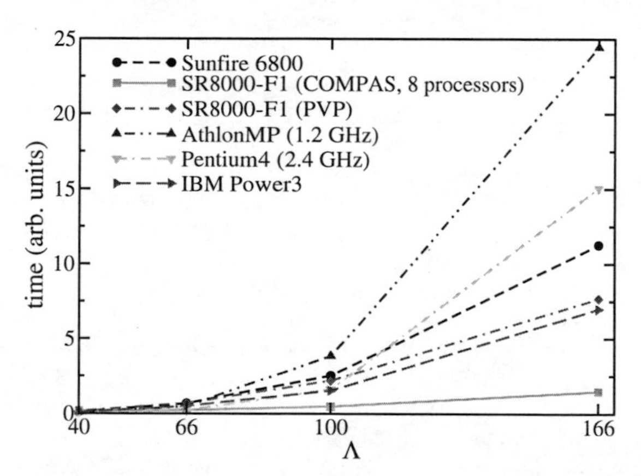

Figure 2.4: Runtime of the three-band QMC code for different computer architectures and system sizes $\Lambda = 40 - 166$ ($T = 1160 - 300$ K); lines are guides to the eye only.

in the figures include typical personal computers (e.g., Pentium4, AthlonMP), compute servers (Power3) and high-performance supercomputers (SR8000-F1, Sunfire 6800). From Fig. 2.4, it is obvious that for large system size ($\Lambda = 166$), only the supercomputers and the Power3 machine are an option due to their large processor caches and their fast memory subsystems. The normal PCs scale

[15] For SR8000-F1 (COMPAS), the data is given for one node which consists of 8 processors.

badly for larger system sizes where the code cannot be executed in the cache any more. The SR8000-F1 in the vectorizing mode (COMPAS)[16] shows nearly an optimal (linear) scaling for increasing system size. Per (single) processor, however, the Power3 and the SR8000-F1 in its single processor mode (PVP) have the best performance for large systems. One has to keep in mind that the speedup (see below) for using more than one processor in parallel is less than linear, hence the best performance for a parallel execution of the code is still attained with SR8000-F1(COMPAS). For smaller systems with $\Lambda \lesssim 100$, the

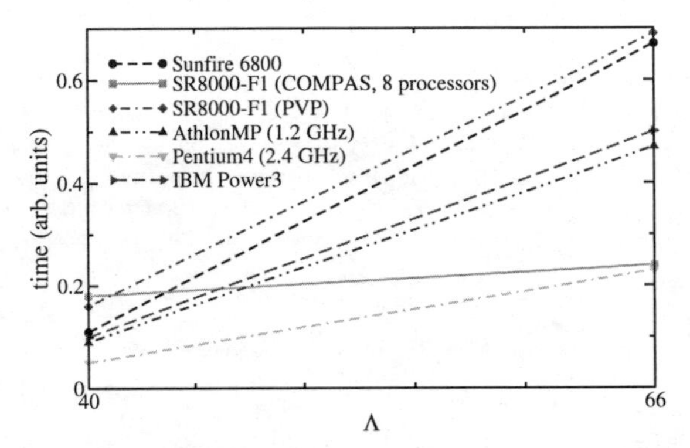

Figure 2.5: Runtime of the three-band QMC code for different computer architectures for system sizes $\Lambda = 40$ ($T = 1160$ K) and $\Lambda = 66$ ($T = 700$ K); lines are guides to the eye only.

Pentium4 with its considerable larger processor frequency is the most efficient choice. This is especially clear from Fig. 2.5, where for small systems, the Pentium4 can even outperform the SR8000-F1(COMPAS) with its 8 processors. For $\Lambda = 40$, the SR8000-F1(COMPAS) is the slowest architecture. At that system size, the AthlonMP which is the slowest processor for large system sizes is quite competitive. As the performance measurements presented here have shown, the choice of the most efficient architecture is strongly dependent on the system size, so a thorough evaluation is necessary before doing extensive production runs of the QMC code.

A second aspect of the performance of the QMC code is the scaling with the number of processors. In Fig. 2.6, the execution time of the parallel three-band QMC depending on the number of processors on the Cray T3E is shown for different systems sizes. It is obvious that the use of more processors reduces the overall computing time considerably. But it can also be noted that the scaling

[16] In this mode, the 8 processors of one node are working together similarly to a vector processor.

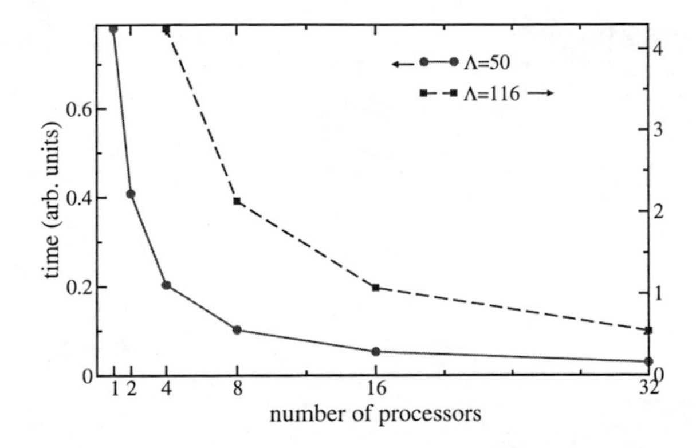

Figure 2.6: Runtime of the three-band QMC code for different number of processors on the Cray T3E for system sizes $\Lambda = 50$ and $\Lambda = 116$; lines are guides to the eye only. Note the different scaling of the time-axis for different system size.

is not ideal, i.e. the runtime is not halved when the number of processors is doubled. A quantity that gives a direct measure of how much faster the parallel algorithm is compared to the sequential one is the speedup S. It is defined as the quotient of the computation time for a single processor T_1 and for p processors T_p, i.e.

$$S(p) = \frac{T_1}{T_p}. \tag{2.32}$$

Fig. 2.7 shows the speedup for the example of Fig. 2.6. For small system size $\Lambda = 50$, the speedup for 32 processors is already reduced considerably compared with a linear (ideal) speedup, it is roughly 27. The larger system, $\Lambda = 116$, has a more favorable speedup of more than 30. If one calculates the efficiency $S(p)/p$, it is about 0.84 for the small and 0.95 for the larger system. The reduction of the efficiency compared to a system with ideal speedup is due to the communication overhead and the sequential portions of the QMC code, i.e., the self-consistency loop and especially the warm-up sweeps.[17] The efficiency is smaller for $\Lambda = 50$ because the communication overhead is larger compared to the computation time.

As the performance measurements have shown, the choice of the most efficient architecture is strongly dependent on the system size, so a thorough evaluation is necessary before doing extensive production runs of the QMC code. Furthermore, a large number of processors can only be utilized effectively for larger system sizes, i.e. for lower temperatures. For small systems, the use of the sequential code is

[17] Although the warm-up sweeps are done in parallel for each process, they are not scaling with the number of processors and thus have to be viewed as a sequential part of the code.

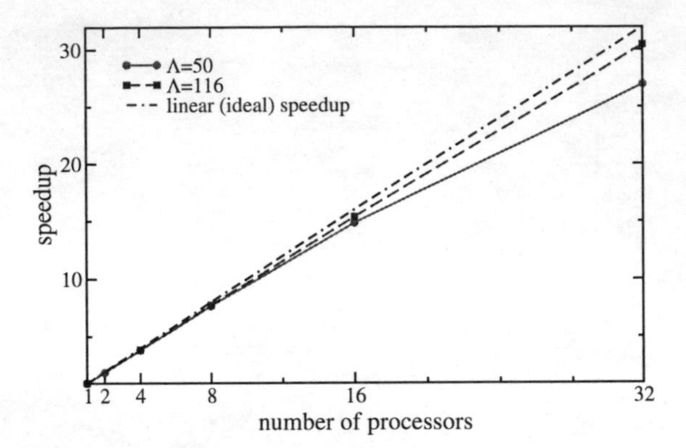

Figure 2.7: Speedup of the three-band QMC code for different number of processors on the Cray T3E for system sizes $\Lambda = 50$ and $\Lambda = 116$ in comparison with linear Speedup; lines are guides to the eye only.

more favorable. In order to increase the efficiency of the parallel code, a fine-grain parallelization would be needed which would also allow for the parallel execution of the warm-up sweeps.

3. V_2O_3 - A SYSTEM ON THE VERGE OF A METAL-INSULATOR TRANSITION

The nature of the Mott-Hubbard metal-insulator transition, i.e., the phase transition between a paramagnetic metal and a paramagnetic insulator that is caused by electronic correlations, is one of the fundamental theoretical problems in condensed matter physics (Mott 1968, Mott 1990, Gebhard 1997). Materials which exhibit such correlation-induced transitions are, e.g., transition metal oxides with partially filled bands near the Fermi edge. V_2O_3 doped with chromium is the most famous example of this class of materials and one of the most intensively studied transition metal oxide compounds (Rice and McWhan 1970, McWhan and Remeika 1970, McWhan, Menth, Remeika, Brinkman and Rice 1973). Fig. 3.1 shows the phase diagram of V_2O_3.

At room temperature, stoichiometric V_2O_3 is paramagnetic and metallic (PM) and crystallizes in the corundum structure. Upon cooling below approximately 150 K, the system undergoes a metal-insulator transition (MIT). This first order phase transition to an insulating antiferromagnetic system (AFI) is accompanied by a reduction of the crystal symmetry to monoclinic structure. It was discovered experimentally in 1946 by Foex, who found a dramatic drop in the electric conductivity by about seven orders of magnitude with a hysteresis of about 10 K, and was later on confirmed for single crystal samples by Morin (1959). By increasing the pressure or doping with Ti, the antiferromagnetic phase boundary is shifted to lower temperatures, the AFI is completely suppressed for pressures above 22 kbar or Ti concentrations above 4.5%. The interpretation of the PM to AFI transition in the context of a Mott-Hubbard transition is difficult due to the coinciding structural and magnetic transition. More interesting from a theoretical point of view is the MIT in the paramagnetic phase, i.e., the first order transition from the paramagnetic metallic to the paramagnetic insulating (PI) phase that appears upon doping with Cr. Among the transition metal oxides, only V_2O_3 shows such a PM to PI transition that doesn't change the crystal structure. Since it is iso-structural (only the c/a ratio changes discontinuously) and the transition line has a anomalous temperature dependence similar to the fluid-solid transition in ^3He (Vollhardt and Wölfle 1990), a predominant electronic origin of this transition is plausible and hence it is considered the classic example of a Mott-Hubbard MIT. However, a recent experimental x-ray absorp-

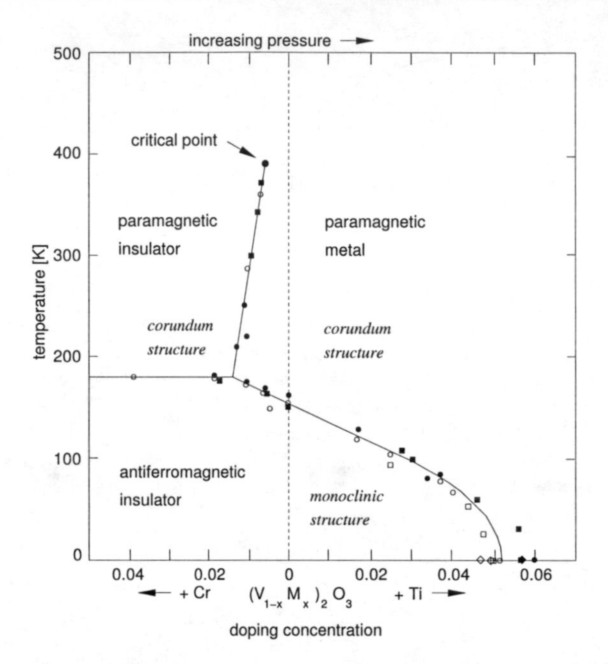

Figure 3.1: Phase diagram of V_2O_3 showing the MIT as a function of pressure and of doping with Cr and Ti; data points from McWhan et al. (1973).

tion study by Pfalzer, Will, Nateprov Jr., Klemm, Eyert, Horn, Frenkel, Calvin and denBoer (2002) for Al-doped V_2O_3 has shown that the PI and the AFI phase have a similar monoclinic distortion, albeit only on a local scale for the PI phase. Furthermore, when going from the metallic to the insulating paramagnetic phase, (local) monoclinic distortions of the crystal structure are detected even before the MIT occurs, i.e., still in the metallic phase. Thus, structural changes may play a role also for the PM to PI transition in V_2O_3, but at present it is not clear to what extent. Further investigations are necessary to clarify this aspect.

In Fig. 3.2, the resistivity of a Cr-doped (paramagnetic) V_2O_3 sample in dependence on the pressure is shown for various temperatures (Jayaraman, McWhan, Remeika and Dernier 1970). The abrupt drop in resistivity at the metal-insulator transition that is about two orders of magnitude at room temperature decreases for higher temperatures and vanishes at about 390 K. Furthermore, the MIT occurs at higher pressure for increasing temperatures, i.e., the temperature dependence of the transition line is anomalous as depicted in the phase diagram (Fig. 3.1).

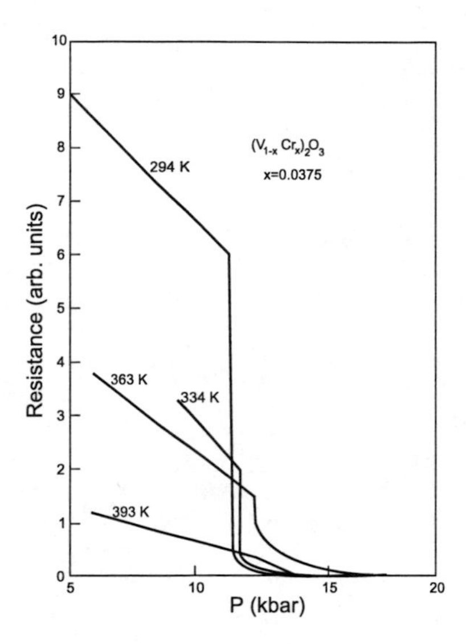

Figure 3.2: Resistivity of V_2O_3 as a function of pressure; from Jayaraman et al. (1970).

An MIT induced by electronic correlations can either be investigated by studying a simplified electronic many-body model that hopefully incorporates some of the basic features of the MIT, or by employing a material-specific approach such as the density functional theory in the local density approximation (LDA). However, conventional band structure calculations yield metallic spectra for correlated systems as V_2O_3, in contrast to experiments. On the other hand, even the simplest possible model, the half-filled, spin $S = 1/2$, single band Hubbard model (Hubbard 1963, Kanamori 1963, Gutzwiller 1963) can give some insight into the Mott-Hubbard transition. Already in the early work of Hubbard (Hubbard 1963, Hubbard 1964a, Hubbard 1964b), the existence of an MIT in the half-filled Hubbard model had been studied. While the insulating phase is described rather well in the Hubbard I and III approximations employed in those investigations, they fail for the metallic, Fermi liquid phase. In the Gutzwiller approximation, the break-down of the Fermi liquid, which is indicated by the collapse of the quasiparticle peak and the divergence of the effective mass, at a critical value of the Coulomb interaction U_c (Brinkman-Rice transition) can be described (Brinkman and Rice 1970), but it cannot reproduce the Hubbard bands which are essential both in the insulating and the strongly correlated metallic phase. Due to the limitations of those approximations, the details of the Mott-Hubbard transition remained unexplained, only the one-dimensional case (Lieb and Wu 1968), which is special since it describes an insulator for all $U > 0$, could be solved.

The development of the dynamical mean-field theory (DMFT, see Sec. 1.2.2) during the last years has considerably advanced our understanding of the metal-insulator transition in the Hubbard model. In DMFT, it is possible to explain the incoherent high-energy features (Hubbard (1964b) bands) and the coherent quasiparticle peak at low energies (Gutzwiller 1963, Brinkman and Rice 1970) in a single framework. Additionally, the transition from the metallic into the insulating phase at $T = 0$, which is accompanied by the collapse of the quasi-particle peak and the break-down of the Fermi liquid phase for Coulomb interaction $U \to U_c^-$, can be described (Georges et al. 1996, Moeller, Si, Kotliar, Rozenberg and Fisher 1995, Rozenberg, Chitra and Kotliar 1999, Bulla 1999). Furthermore, the coexistence of the metallic and insulating phase below a critical point at temperature T_c and the existence of a first-order transition were calculated in DMFT with a multitude of methods (Georges and Krauth 1992, Rozenberg et al. 1992, Bulla 1999, Rozenberg et al. 1999, Joo and Oudovenko 2001, Blümer 2002), in agreement with the experimental results for V_2O_3. Rozenberg, Kotliar, Kajüter, Thomas, Rapkine, Honig and Metcalf (1995) investigated the MIT in V_2O_3 within the DMFT in the one-band Hubbard model. Later studies took into account the influence of orbital degeneracy within a two- (Rozenberg 1997, Han, Jarrell and Cox 1998, Held and Vollhardt 1998) and three-band (Han et al. 1998) Hubbard model for the Bethe lattice, i.e. for a semi-circular density of states. Recently, it was shown in a detailed analysis of the conductivity (Limelette, Georges, Jérome, Wzietek, Metcalf and Honig 2003) that the critical exponents are of the liquid-gas transition type[1] except for a narrow region around the critical point.

Although certain basic features of the Mott-Hubbard MIT and the phase diagram of V_2O_3 can be explained with the Hubbard model in DMFT, a detailed, material specific description is not possible. In our work presented in this chapter, we therefore applied the LDA+DMFT scheme to study the metal-insulator transition in V_2O_3. The details of this scheme are discussed in chapter 1. Another LDA+DMFT investigation on the same subject was performed by Laad, Craco and Müller-Hartmann (2003) who employed iterative perturbation theory to solve the DMFT equations.

[1] This agrees with a Landau theory for the Mott-Hubbard transition within DMFT (Moeller et al. 1995, Kotliar 1999, Kotliar, Lange and Rozenberg 2000).

3.1 The crystal structure of V_2O_3

In the paramagnetic metallic phase, V_2O_3 has a trigonal corundum structure with space group $R\bar{3}c$ (D_{3d}^6). In the often used hexagonal notation, the lattice constants are $a = 4.9515$Å and $c = 14.003$Å (Dernier 1970). The structure is based on a hexagonally closest packing of the oxygen atoms with ABAB layer sequence in the c direction. Two thirds of the centers of the octahedra thus formed by the oxygen atoms are occupied by vanadium atoms, leaving every third position in each crystal direction vacant (Mattheiss 1994). This results in a vanadium subgrid with ABCABC layer sequence, so that the elementary cell has six layers in the c direction. Fig. 3.3 shows the crystal structure with the occupied oxygen octahedra marked with grey faces and the unoccupied octahedra with light grey lines. Every vanadium atom has four nearest neighbors, three in the basal plane (sharing octahedral edges) and one in c direction (sharing octahedral faces). Due to their repulsion, the neighboring vanadium atoms are shifted slightly in

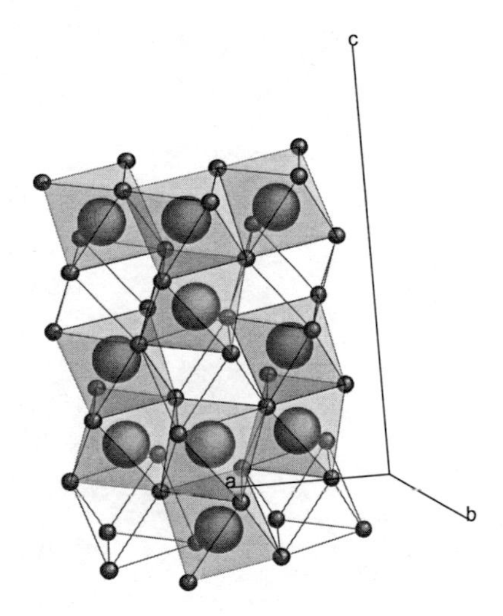

Figure 3.3: Crystal structure of V_2O_3. For clarity, only a part of the hexagonal unit cell is shown; sizes of the atoms are not true to scale. Small spheres: oxygen atoms; large spheres: vanadium atoms; light grey faces: oxygen octahedra occupied by vanadium atoms; light grey lines: unoccupied oxygen octahedra.

the direction of the vacant octahedral positions, but still form "pairs" along the hexagonal c-axis. The oxygen octahedra are distorted also which can be seen in the marked deviation of the internal parameters for the Wyckoff positions. Instead of (0,0,1/3) for vanadium (position (12c)) and (1/3,0,1/4) for oxygen (position (18e)) in the ideal hexagonal arrangement, they are (0,0,0.34630) and (0.31164,0,1/4) in V_2O_3, respectively (Dernier 1970).

In the chromium doped paramagnetic insulating phase, the corundum crystal symmetry is preserved but the structure parameters are changed by the substitution with the isovalent but smaller Cr atoms. For the compound with 3.8% Cr-doping used in most of our calculations, the positional parameters are (0,0,0.34870) for vanadium and (0.310745,0,1/4) for oxygen, the lattice constants are $a = 4.9985$Å and $c = 13.912$Å (Dernier 1970). Fig. 3.4 illustrates the distinct displacements between the insulating and metallic phase. Although the lattice

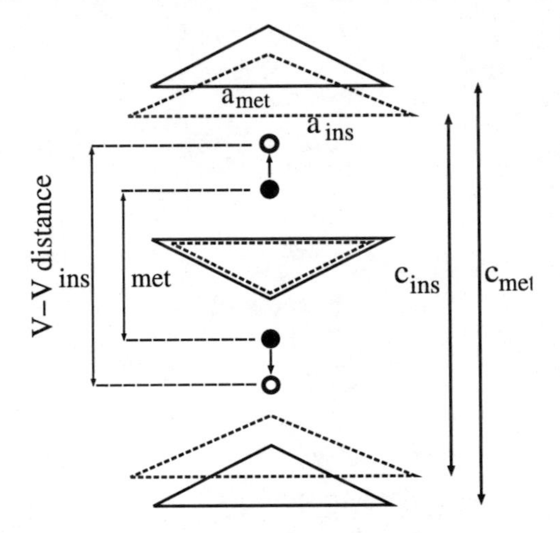

Figure 3.4: Illustration of the displacement of the vanadium and oxygen atoms in Cr-doped V_2O_3 after Dernier (1970) (filled circles denote vanadium in the metallic, open circles in the insulating phase).

constant of the c-axis decreases in the insulator, the V-V distance in the vanadium pairs is increased. The octahedral faces between the vanadium pairs shrink (middle triangles in Fig. 3.4), whereas the octahedral faces pointing towards the vanadium vacancies (upper and lower triangles in Fig. 3.4) are enlarged, leading to an increased a-axis lattice constant in $(V_{1-x}Cr_x)_2O_3$. Therefore all the nearest-neighbor vanadium distances, in the ab-plane as well as in c-direction, are enlarged in the insulating phase which should lead to a reduction in the bandwidth of the t_{2g}-bands. The vanadium-oxygen distances remain essentially the same. In his investigation of V_2O_3 and $(V_{1-x}Cr_x)_2O_3$ and the comparison with other α corundum oxides, Dernier (1970) concluded that the metallic properties of pure V_2O_3 are not only due to hopping processes between the vanadium pairs in the c-direction, but also due to hopping within the ab-plane.

Entering the antiferromagnetic insulating phase, the vanadium distances are further enlarged and the V-V pairs are tilted in the hexagonal (1000) plane by 1.8° in relation to the c-axis (Brückner, Oppermann, Reichelt, Terukow, Tschudnowski and Wolf 1983). While the position of the oxygen atoms is nearly unchanged, the abovementioned tilting reduces the symmetry to a monoclinic structure C 2/c (space group 15, point group 2/m). The volume increases by 1% which explains why the antiferromagnetic phase is suppressed at high pressures.

3.2 Electronic structure and band structure calculations

In the octahedral oxygen coordination of V_2O_3, the cubic crystal field leads to a splitting of the five degenerate atomic V-3d bands into three degenerate t_{2g}-bands and two degenerate e_g^σ-bands (Cox 1992). Due to the trigonal distortion of the oxygen octahedra, the t_{2g}-bands are further split up into one a_{1g}-band and two degenerate e_g^π-bands,[2]. Both the a_{1g}- and the e_g^π-bands are to a certain extent split into bonding and antibonding contributions (see Fig. 3.5). Goodenough (1970, 1971) proposed a model for the qualitative description of

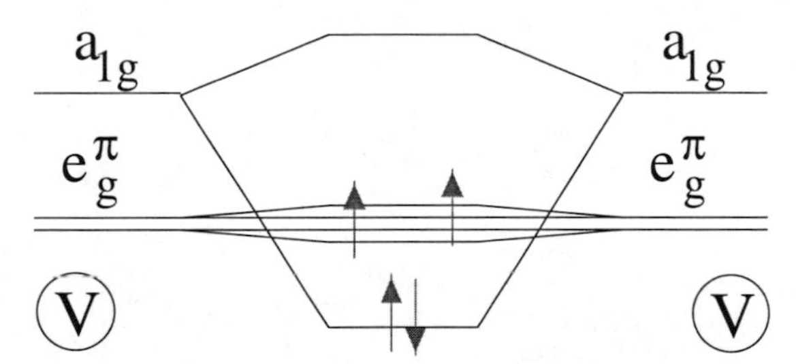

Figure 3.5: Left and right: splitting of the t_{2g}-orbitals of the vanadium atoms in the corundum crystal structure according to Castellani et al. (1978). Middle: Formation of a chemical bond for a single V-V pair along the c-axis.

the metal-insulator transition with the two V-3d electrons situated mainly in the

[2] The designations t_{2g}, e_g and a_{1g} originate from group theory (see Hammermesh (1989) for details) the indices π and σ indicate the type of bonding between the O-2p and V-3d orbitals involved.

bonding e_g^π-bands. The isotropic metallic properties are explained by a small overlap with the bonding a_{1g}-band which is at higher energies due to the large c/a ratio. Another quite contrary model was proposed by Castellani et al. (1978). As mentioned before, V-V pairs are formed in the *c*-direction of the corundum structure. The a_{1g}-orbitals which are directed along the *c*-axis mediate a strong hybridization between the V atoms in the pairs. The importance of those intra-pair interactions for the interpretation of spectroscopy results of pure and Cr-doped V_2O_3 was already emphasized by Allen (1976). In the Castellani picture, the strong hybridization leads to a bonding a_{1g}-band below the e_g^π-bands, which is fully occupied with two electrons (see Fig. 3.5). The remaining two electrons (one per V site) in the two degenerate, partially filled e_g^π-bands lead to a orbitally degenerate spin-$\frac{1}{2}$ state for each V site with a complicated orbital and spin ordering pattern that accounts for the unusual properties of the antiferromagnetic phase. For a long time the spin-$\frac{1}{2}$ model by Castellani et al. (1978) was the standard model for V_2O_3. In recent years, new theoretical and experimental results have been challenging this model. Park, Tjeng, Tanaka, Allen, Chen, Metcalf, Honig, de Groot and Sawatzky (2000) concluded from the polarization dependence of their x-ray absorption measurements that the system is in a spin-1 state. They further found a predominant $e_g^\pi e_g^\pi$ ground state with a small admixture of $a_{1g}e_g^\pi$ configurations. The same result was obtained by Ezhov, Anisimov, Khomskii and Sawatzky (1999) in their LDA+U calculations and also in our LDA+DMFT calculations (see Sec. 3.3). Nevertheless, many theoretical studies of V_2O_3 are still based on the idea of a strong hybridization in the V-V pair with the intra-pair a_{1g} hopping as the main contribution and the inter-pair hopping as a perturbation (Mila, Shiina, Zhang, Joshi, Ma, Anisimov and Rice 2000, Tanaka 2002, Di Matteo, Perkins and Natoli 2002). In those model calculations, the values of the hopping parameters were obtained by a least-square fit of the LDA bands to a model Hamiltonian with nearest neighbor hopping. Taking into account next-nearest neighbor hoppings (which were found to be non-negligible) in the fit to a model Hamiltonian, Elfimov, Saha-Dasgupta and Korotin (2003) found a significantly reduced a_{1g}-a_{1g} hopping in the vanadium pair. The inter-pair hopping can therefore not be considered as a small contribution and cannot be taken as a perturbation in theoretical calculations.

In order to perform LDA+DMFT computations, band structure calculations (Hohenberg and Kohn 1964, Kohn and Sham 1965) in the local density approximation (LDA) within the augmented spherical wave (ASW) method (Williams, Kübler and Gelatt Jr. 1979, Eyert 2000b) were done by Eyert (2000a). The resulting band structures along selected high-symmetry lines in the first Brillouin zone of the hexagonal lattice (see Fig. 3.6) are shown in Figs. 3.7 and 3.8 for V_2O_3 and $(V_{0.962}Cr_{0.038})_2O_3$, respectively, the corresponding densities of states in Figs. 3.9 and 3.10. The length of the bars in Figs. 3.7 and 3.8 is a measure of the contribution of the a_{1g}-orbitals to the respective wave functions.

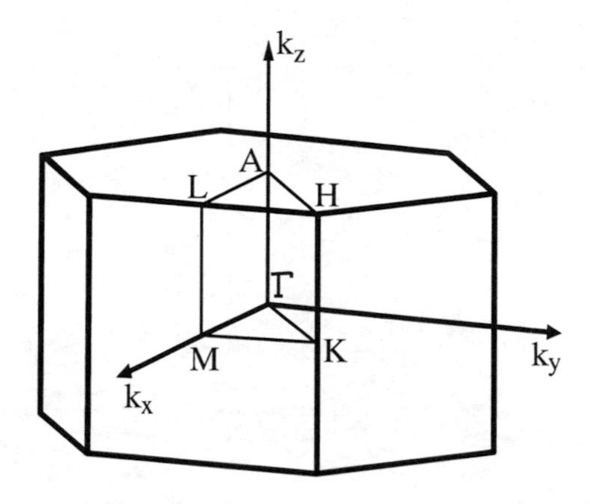

Figure 3.6: First Brillouin zone of the hexagonal lattice.

Overall, our results are in good agreement with those published by Mattheiss (1994). The oxygen bands are in the range between -9 and -4 eV (not show in the figures), the vanadium bands are split up due to the octahedral oxygen surrounding into the partially filled t_{2g}-bands (with two electrons per V atom) between -1 and 1.5 eV and empty e_g^σ-bands between 2 and 4 eV. The splitting of the t_{2g}-band due to the trigonal distortion can be clearly seen in Figs. 3.9 and 3.10. The absolute value of the splitting, which is ≈ 0.3 eV for the centers of gravity, is much smaller than the t_{2g}-bandwidth (≈ 2 eV). Nevertheless, since the value of the Coulomb interaction parameter U is larger than the bandwidth ($U \sim 5$ eV), the small trigonal splitting strongly influences the orbital ground state of the vanadium atom obtained from LDA+DMFT calculations (see Sec. 3.3). The shape of the a_{1g} DOS is markedly different from that of the e_g^π-bands, the weight is shifted from the center of the band to the band edges. This is in accordance to the picture of an a_{1g}-band that is split into a bonding and an antibonding band due to the strong hybridization in the V-V pair. As will be shown later, this splitting is smeared out by the correlations introduced in the DMFT calculations. The LDA results for the metallic phase (V_2O_3) and the insulating phase ($(V_{0.962}Cr_{0.038})_2O_3$) are qualitatively identical, there is no gap in the results of the calculation for the insulating crystal structure. We observe a narrowing of the t_{2g}- and e_g^σ-bands of ≈ 0.2 eV and ≈ 0.1 eV, respectively, and a slight downshift of the centers of gravity of the e_g^π-bands.

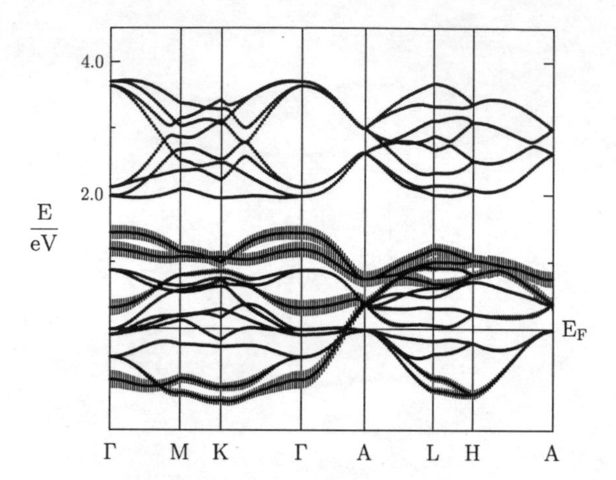

Figure 3.7: Electronic bands of V_2O_3 along selected symmetry lines within the first Brillouin zone of the hexagonal lattice, Fig. 3.6. The width of the bars given for each band indicates the contribution from the a_{1g}-orbitals.

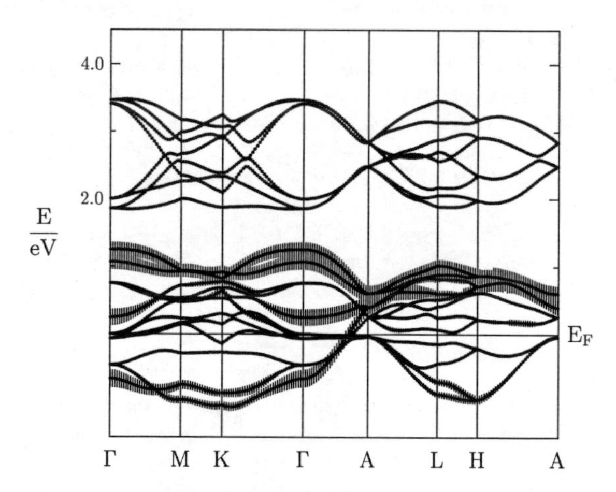

Figure 3.8: Electronic bands of $(V_{0.962}Cr_{0.038})_2O_3$ (same as Fig. 3.7 for Cr-doped V_2O_3).

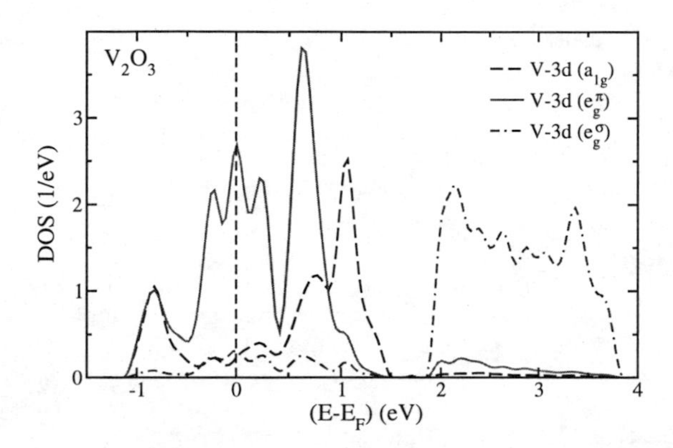

Figure 3.9: Partial V-3d densities of states (DOS) of V_2O_3 per unit cell.

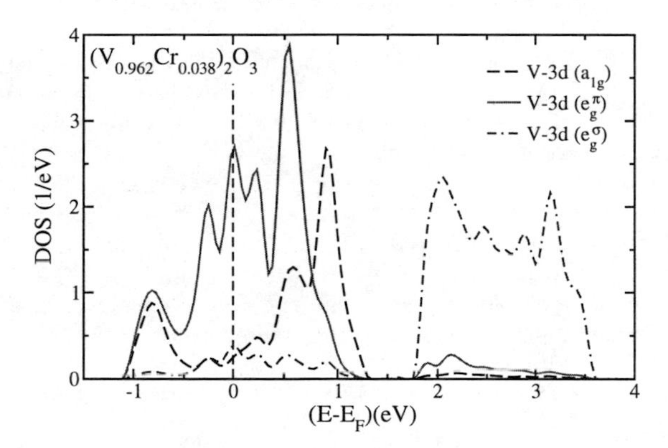

Figure 3.10: Partial V-3d densities of states (DOS) of $(V_{0.962}Cr_{0.038})_2O_3$ per unit cell.

3.3 LDA+DMFT spectra

With the LDA densities of states from Figs. 3.9 and 3.10 as input, we performed DMFT(QMC) calculations (as described in chapter 1) for one a_{1g}- and two degenerate e_g^π-bands. The oxygen bands (which are far below the Fermi edge) and the e_g^σ-bands (which are above the Fermi energy, separated by a gap of ≈ 1 eV from the t_{2g}-bands and completely empty) were not taken into account. While the Hund's rule coupling constant J can be obtained from constrained LDA with good accuracy (it is insensitive to screening effects), the Coulomb repulsion calculated in constrained LDA has a typical uncertainty of about 0.5 eV (Nekrasov et al. 2000). We therefore used $J = 0.93$ eV from Solovyev et al. (1996) and adjusted U to get a metallic solution for the crystal structure of pure V_2O_3 and an insulating solution for the crystal structure of $(V_{0.962}Cr_{0.038})_2O_3$.[3] The resulting spectra for the t_{2g}-bands (i.e. summed over a_{1g}- and e_g^π-bands) at different values of U are shown in Fig. 3.11. At $U = 4.5$ eV, we observe metallic behavior for both crystal structures with a lower Hubbard band at about 1 eV, an upper Hubbard band split by Hund's rule coupling with peaks at 1 eV and 4 eV and a quasiparticle peak at the Fermi edge. By contrast, the spectra at $U = 5.5$ eV both show nearly insulating behavior. The quasiparticle peak vanishes and a pseudo-gap is formed, its spectral weight is transfered into the thereby strongly enhanced lower Hubbard band. In the two-peak structure above the Fermi edge, only minor changes are visible. At the intermediate value $U = 5$ eV, we observe qualitatively different spectra for the two crystal structures. Whereas pure V_2O_3 exhibits a small peak at the Fermi edge (which is the residue of the quasiparticle peak at lower U) and is therefore metallic, the spectrum for the Cr-doped structure shows a pronounced minimum which indicates that the system is nearly insulating. The observed smooth crossover between a metal-like and an insulator-like behavior of the spectra (instead of a sharp metal-insulator transition as expected for temperatures below the critical point, i.e. for $T < 400$ K in experiment) is due to the high temperature of $T = 0.1$ eV ≈ 1160 K of our QMC calculations. Our critical value of U of about 5 eV for the MIT is in agreement with constrained LDA calculations by Solovyev et al. (1996) who obtained a slightly smaller value than 5 eV for the t_{2g}-orbitals by analyzing the charging energy between di- and trivalent vanadium ions in the octahedral environment of $LaVO_3$. It is also in accordance to fits of spectroscopy data for vanadium oxides to model calculations which yielded $U = 4 - 5$ eV.

In order to get spectra at experimentally relevant temperatures, we performed calculations at $T = 700$ K and $T = 300$ K. As the computation time of the QMC algorithm is proportional to T^{-3} (for details, see Sec. 2.1), those calculations were computationally very expensive and were performed on a supercomputer.[4]

[3] The inter-band Coulomb repulsion $U' = U - 2J$ is fixed by orbital rotational symmetry.

[4] We used the Hitachi SR8000-F1 in the Leibniz-Rechenzentrum in Munich and the IBM

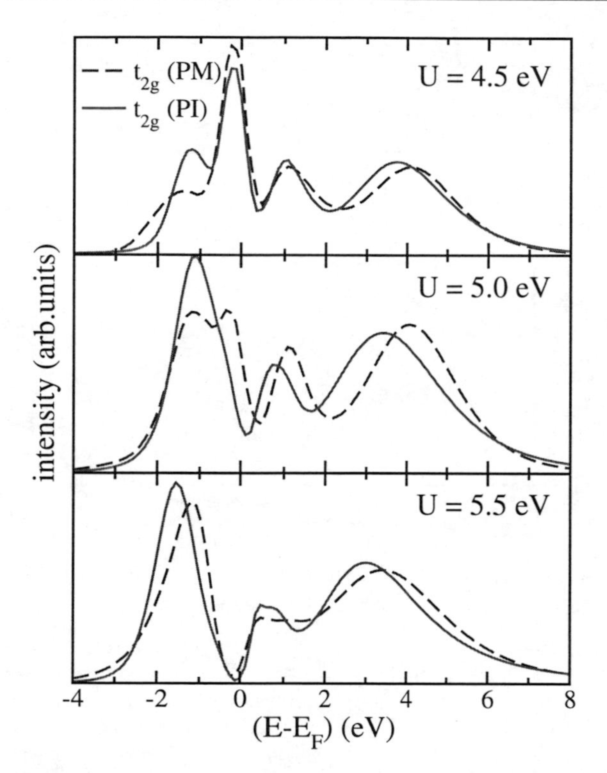

Figure 3.11: LDA+DMFT(QMC) spectra for paramagnetic insulating (PI) $(V_{0.962}Cr_{0.038})_2O_3$ and metallic (PM) V_2O_3 at $U = 4.5$, 5 and 5.5 eV, $T = 1160$ K.

In Fig. 3.12, the spectra for V_2O_3 at $T = 1160$ K, $T = 700$ K and $T = 300$ K and for $(V_{0.962}Cr_{0.038})_2O_3$ at $T = 1160$ K and $T = 700$ K are shown.[5] The insulating spectra show only small changes for lower temperatures. In the lower Hubbard band, some weight is shifted from the region near the Fermi edge to the peak at 1 eV, whereas the spectral intensity of the minimum at 0 eV is not affected. In the metallic phase, the lower and upper Hubbard bands are only minimally changed for lower temperatures. However, the quasiparticle peak becomes sharper and much more pronounced. This is due to the smoothing of the Abrikosov-Suhl

p690 e-server cluster 1600 in the Forschungszentrum Jülich (Jump) in our calculations.

[5] The calculations at $T = 300$ K for the insulating crystal structure did not yield sufficiently accurate results and are therefore omitted in Fig. 3.12.

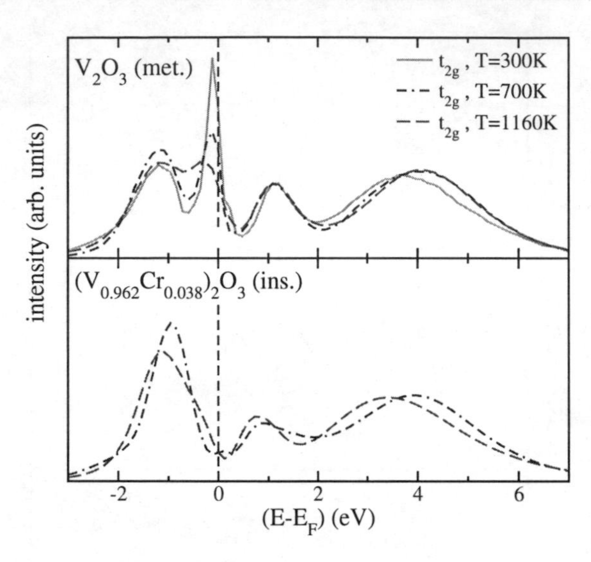

Figure 3.12: LDA+DMFT(QMC) spectra for paramagnetic insulating $(V_{0.962}Cr_{0.038})_2O_3$ and metallic V_2O_3 at $U = 5$ eV.

quasiparticle resonance at temperatures above the Kondo temperature. A similar behavior is observed in the Anderson impurity model (Hewson 1993), but here it occurs at considerably lower temperatures which is apparently an effect of the DMFT self-consistency cycle. Since the t_{2g}-bands are not degenerate, it is interesting to take a look at the band-resolved spectra. In Fig. 3.13, the a_{1g}- and and one e_g^π-band for $U = 5.0$ eV for both input densities of states are shown at 700 K. Obviously, the two types of bands have very different contributions to the overall spectrum. The lower Hubbard band and the quasiparticle peak are dominated by the e_g^π-bands.[6] The two electrons per vanadium site will therefore mainly occupy the e_g^π-bands with only a small admixture of a_{1g}-occupation (for details, see Sec. 3.4). In the upper Hubbard band, the a_{1g}-orbital is mainly responsible for the peak at 1 eV, whereas the contributions are about equal for the peak at higher energies. To illustrate the origin of the basic features of the spectrum, the a_{1g}- and one e_g^π-band for $U = 5.5$ eV at $T = 1160$ K are displayed in Fig. 3.14 for $(V_{0.962}Cr_{0.038})_2O_3$. As we will show in Sec. 3.4, the predominant local configuration is an $e_g^\pi e_g^\pi$ spin-1 configuration, i.e. two spin-aligned electrons in the e_g^π-orbitals, with a small admixture of $a_{1g}e_g^\pi$ spin-1 configurations. In the following, we will examine the spectrum of the atomic Hamiltonian, which is a

[6] The e_g^π contribution in Fig. 3.13 has to be doubled since only one of the two degenerate bands is shown.

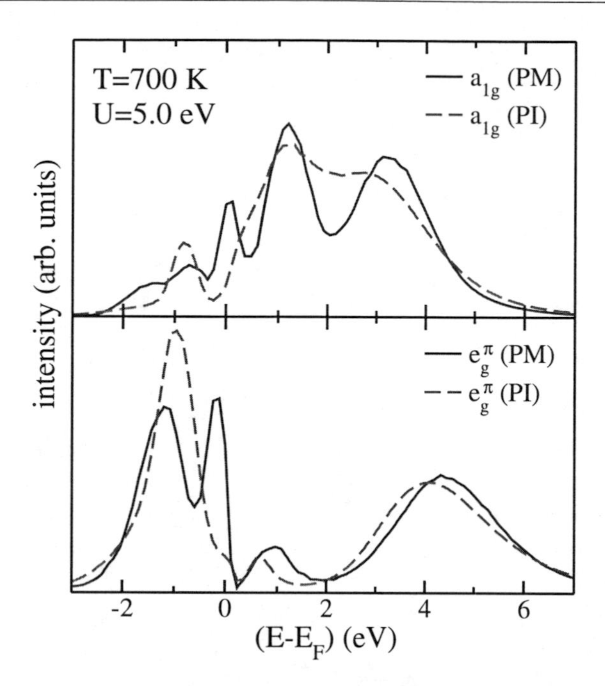

Figure 3.13: Band resolved comparison of LDA+DMFT(QMC) spectra for paramagnetic insulating (PI) $(V_{0.962}Cr_{0.038})_2O_3$ and metallic (PM) V_2O_3 at $U = 5$ and $T = 700$ K.

reasonable simplification for the insulating phase. Removing one electron from the $e_g^\pi e_g^\pi$ spin-1 configuration, i.e. a $e_g^\pi e_g^\pi \rightarrow e_g^\pi$ transition, results in an energy gain of $U' - J - \mu \approx -1.5$ eV, which is in good agreement with the position of the lower Hubbard band. The same applies to the transition $a_{1g}e_g^\pi \rightarrow a_{1g}$, only the transition $a_{1g}e_g^\pi \rightarrow e_g^\pi$ has a slightly higher energy gain due to the trigonal splitting between a_{1g}- and e_g^π-bands. The main contributions to the upper Hubbard e_g^π-band are the $e_g^\pi e_g^\pi \rightarrow e_g^\pi e_g^\pi e_g^\pi$ transitions. Since the third e_g^π-electron cannot be spin-aligned, the transition costs $U + U' - \mu = U + J + (U' - J - \mu) \approx 4.4$ eV, which roughly agrees with the position of the upper peak of the upper Hubbard band. Adding one a_{1g} electron instead ($e_g^\pi e_g^\pi \rightarrow e_g^\pi e_g^\pi a_{1g}$), there are two configurations possible, one with aligned spins ($2U' - 2J - \mu \approx 0.7$ eV) and one without spin-alignment ($2U' - \mu \approx 2.6$ eV). The Hund's rule coupling leads to a splitting of the upper Hubbard a_{1g}-band, this is in good agreement with the two-peak structure in Fig. 3.14. Since we have a small admixture of $a_{1g}e_g^\pi$ spin-1 configurations, the upper Hubbard e_g^π-band is also split with a low peak at 0.5 eV due to $a_{1g}e_g^\pi \rightarrow$

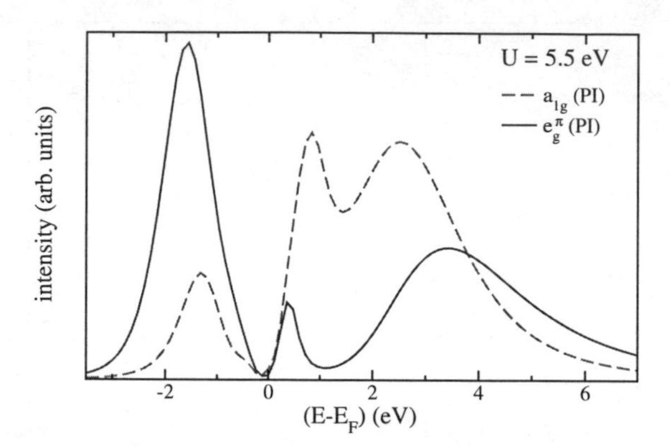

Figure 3.14: LDA+DMFT(QMC) spectrum (a_{1g}- and one e_g^π-band) for paramagnetic insulating $(V_{0.962}Cr_{0.038})_2O_3$ (PI) for $U = 5.5$ eV; $T = 1160$ K.

$a_{1g}e_g^\pi e_g^\pi$ transitions. Because of the splitting of the upper Hubbard band, spectral weight is shifted towards the Fermi edge. Therefore the insulating gap becomes very small compared to a gap of order $U' \approx 3$ eV which one would expect in a one-band Hubbard model. This Hund's rule splitting also explains the puzzle of the unrealistically small Coulomb repulsion (about 1 eV) and bandwidth (less than 0.5 eV) found by Rozenberg et al. (1995) in their attempt to fit the small experimental optical gap with a one-band Hubbard model.

Since the Hund's rule coupling is obviously of major importance for the spectrum above the Fermi edge,[7] we performed additional high-temperature calculations for reduced coupling $J = 0.7$ eV and $U = 4.5$ eV, keeping the inter-band coupling U' nearly constant. As can be seen in Fig. 3.15, the spectrum below the Fermi edge is hardly affected by the small J value, the physical properties should therefore not change much. We find, for example, that the spin-1 state is almost unaffected by lower J values, even for $J = 0.5$ eV, the local moment $\langle m_z^2 \rangle = 3.85$ is near the maximum value of 4. To recover the spin-$\frac{1}{2}$ picture of Castellani *et al.* would require unrealistically small values for the Hund's rule coupling in our calculations.[8] Above the Fermi energy, the lower peak is shifted slightly and the upper peak is shifted considerably towards lower energies. In a comparison with experimental data, a smaller Hund's rule coupling would therefore not make much difference for, e.g., photoemission spectra or measurements of the gap at

[7] For systems with less than half-filled bands, for systems above half-filling, e.g. NiO, the situation is reversed and the splitting occurs in the lower Hubbard band.

[8] It would also require a LDA density of states with an a_{1g}-band split into a bonding and antibonding peak of equal weight to obtain an a_{1g}-singlet and an unpaired spin in the e_g^π-bands.

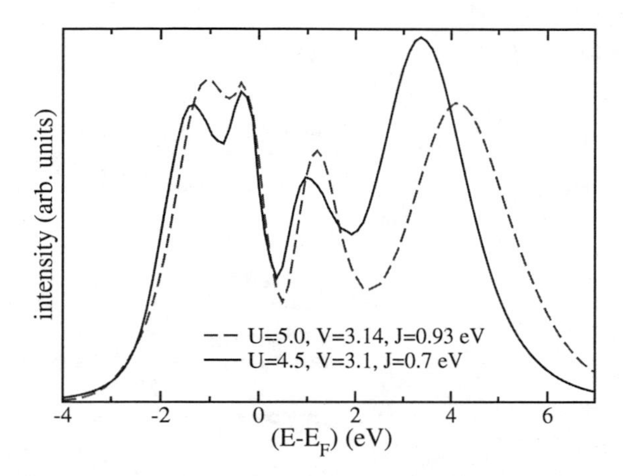

Figure 3.15: Comparison of the LDA+DMFT(QMC) V_2O_3 spectra at two strengths of the exchange interaction ($T = 1160$ K): $J = 0.93$ eV (as obtained from constrained LDA and used in all other figures), $U = 5.0$ eV, $U' = 3.14$ eV, and $J = 0.7$ eV, $U = 4.5$ eV, $U' = 3.1$ eV.

the Fermi edge, but it is of considerable importance for the comparison with XAS data, see Sec. 3.5 for details.

3.4 Changes across the Mott-Hubbard transition

3.4.1 Local magnetic moment and orbital occupation

Besides the spectral densities, the spin and orbital degrees of freedom give valuable insight into the MIT in the paramagnetic phase of V_2O_3. The spin state of the system can be deduced from the squared local magnetic moment $\langle m_z^2 \rangle = \langle (\sum_m [\hat{n}_{m\uparrow} - \hat{n}_{m\downarrow}])^2 \rangle$ which is shown in the upper part of Fig. 3.16. Throughout the transition around $U = 5$ eV and even down to 3 eV, we find a value $\langle m_z^2 \rangle \approx 4$. Below 3 eV, the Hund's rule coupling J has to be reduced to avoid an unphysical attractive Coulomb interaction (namely, an energy gain of $U - 3J < 0$ for the addition of a spin-aligned electron to a singly occupied site) and this reduction of the J value finally leads to a decrease of the squared local moment. The value $\langle m_z^2 \rangle \approx 4$ corresponds to two spin-aligned electrons in the three (a_{1g},e_g^π,e_g^π) orbitals, i.e. a spin-1 state in the vicinity of the metal-insulator transition. This is in good agreement with polarization dependent x-ray measurements of Park et al. (2000) and with high-temperature susceptibility measurements which yielded an effective magnetic moment $\mu_{eff} = 2.66\mu_B$ (Arnold and Mires 1968) close to the $S = 1$ value $\mu_{eff} = 2.83\mu_B$. In contrast to our results of an unchanged spin state throughout the transition, the local magnetic moment changes strongly at the MIT in one-band Hubbard model calculations (Georges et al. 1996), $\langle m_z^2 \rangle$ was even used as an indicator for the transition. The different behavior of the local magnetic moment is simply due to the fact that in an one-band model, a high-spin state is connected to a singly occupied band whereas the low-spin state is only possible for a doubly occupied band. Therefore at large values of U (i.e. in the insulating regime), a singly occupied site and hence a high-spin state is energetically favorable. In a multiband model, the spin state is not connected to the Coulomb interaction U but rather to the Hund's rule coupling J which favors a high-spin state but does not change at the MIT.

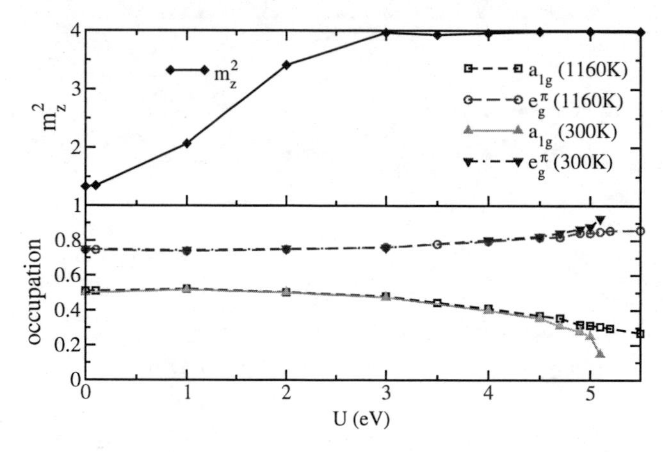

Figure 3.16: Spin (upper figure) and orbital occupation (lower figure) vs. Coulomb interaction U for metallic V_2O_3.

The orbital occupation for metallic V_2O_3 is shown in the lower part of Fig. 3.16. We observe predominantly occupied e_g^π-orbitals, the a_{1g}-orbital is less than half-filled at all values of U. The occupation of the a_{1g}-orbital decreases considerably near the metal-insulator transition, especially for the low-temperature curve ($T = 300$ K). In our high-temperature calculations ($T = 1160$ K) for $U = 5$ eV, we find an occupation of the (a_{1g},e_g^π,e_g^π) orbitals of (0.37,0.815,0.815) for (metallic) V_2O_3 and (0.28,0.86,0.86) for (insulating) $(V_{0.962}Cr_{0.038})_2O_3$. Park et al. (2000) concluded on the basis of an analysis of their linear dichroism data that the ratio between (e_g^π,e_g^π) and (a_{1g},e_g^π) configurations is 1:1 for the metallic phase and 3:2 for the insulating phase which amounts to an electron occupation of (0.5,0.75,0.75) and (0.4,0.8,0.8), respectively. While our results show a higher e_g^π-

and lower a_{1g}-occupation (especially for lower temperatures), the tendency of a reduction of the a_{1g}-admixture in the insulating phase is well reproduced. Another experimental confirmation for an (e_g^π,e_g^π) ground-state configuration comes from polarized neutron defraction measurements in paramagnetic V_2O_3 by Brown, Costa and Ziebeck (1998). In their measurements of the spatial distribution of the field-induced magnetization, they found that the moment induced on the vanadium atoms is mainly due to the electrons in the e_g^π-bands with only a minor contribution from a_{1g}-electrons.

The prevailing e_g^π-occupation which leads to the (e_g^π,e_g^π) ground-state configuration is already found in the LDA results (see Fig. 3.9). Since the center of gravity of the a_{1g}-band is 0.3 eV higher in energy than that of the e_g^π-bands and the a_{1g}-band is asymmetric with only a minor part of the spectral weight lying below the Fermi edge, the occupation is about 0.55 in the a_{1g}-band and 0.72 in each of the e_g^π-bands. The a_{1g}-occupation is further reduced in our LDA+DMFT calculations due to the strong correlations, especially in the insulating phase where the Coulomb interaction ($U > 5$ eV) is considerably larger than the bandwidth ($W \approx 2$ eV). In accordance with our results, Ezhov et al. (1999) found a spin-1 ground state and a (e_g^π,e_g^π) orbital configuration in LDA+U for the antiferromagnetic insulating phase of V_2O_3.

3.4.2 Quasiparticle renormalization and spectral weight at the Fermi level

To get more insight into the metal-insulator transition and more information on the critical U value, we calculated the quasiparticle weight $Z = (1-\partial\Sigma(\omega)/\partial\omega)^{-1}$ via the slope of a third-order polynomial at $\omega = 0$ that is fitted to the imaginary part of the QMC self-energy Im $\Sigma(i\omega_n)$ for small Matsubara frequencies ω_n. Fig. 3.17 shows the resulting quasiparticle weight for the a_{1g}- and e_g^π-bands at 300 K with respect to the Coulomb interaction U. For both types of bands, at small U values Z decreases rapidly with increasing Coulomb interaction, followed by a plateau at medium U values for the e_g^π-bands and a further but smaller decrease for the a_{1g}-band. Although the Z value for the a_{1g}-band is higher than for the e_g^π-bands for all U values, the overall behavior up to $U \approx 5$ eV is quite similar. Near the MIT ($U > 5$ eV) however, the quasiparticle weight for the a_{1g}-band remains constant, whereas that of the e_g^π-bands goes to zero. Considering this puzzling behavior on the basis of a one-band Hubbard model (where the MIT is accompanied by $Z \to 0$, i.e. a diverging effective mass $m^*/m = 1/Z$), one could conclude that the MIT only occurs for the e_g^π-bands. However, the LDA+DMFT spectrum clearly shows insulating behavior for both types of bands (compare Fig. 3.14) at large Coulomb interactions. To resolve this puzzle and to get a

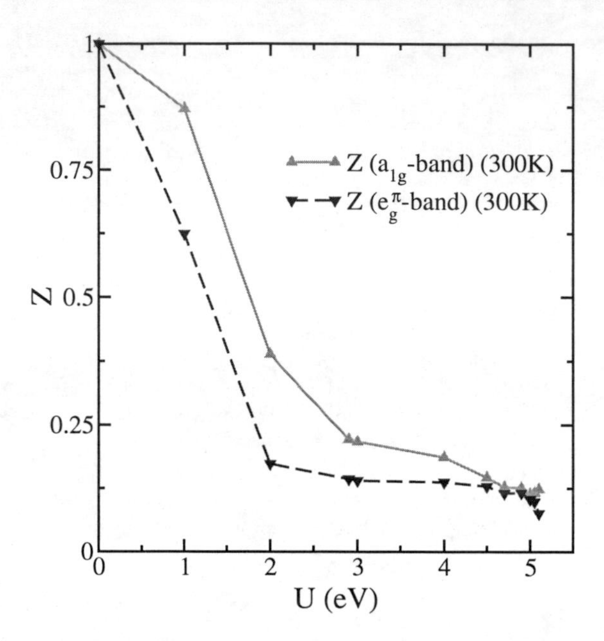

Figure 3.17: Quasiparticle weight Z for the a_{1g} and the e_g^π-bands vs. U, using the crystal structure of metallic V_2O_3.

better understanding of the transition, we calculate

$$-\frac{\beta}{\pi}G(\tau=\beta/2)=\int A(\omega)\underbrace{\frac{\beta}{\pi}\frac{\exp(-\beta/2\omega)}{1+\exp(-\beta\omega)}}_{K(\omega)}\,\mathrm{d}\omega. \tag{3.1}$$

The quantity $-\frac{\beta}{\pi}G(\tau=\beta/2)$ (shown in Fig. 3.18) measures the spectral weight $A(\omega)$ in the region where the kernel $K(\omega)$ contributes, i.e. in an area with a width proportional to $T=1/\beta$ around the Fermi edge. It is therefore ideal to study the vanishing of the spectral weight at the Fermi energy and opening of the gap at the transition. Since it does not depend on the analytical continuation to real frequencies, it can further be used to cross-check our spectra derived with the maximum entropy method. The high-temperature curves (T = 1160 K) in Fig. 3.18 show a smooth decrease of the spectral weight around 5 eV, the MIT is smeared out to a crossover and is only signaled by a change in curvature slightly above 5 eV. However, at 300 K, the spectral weight at the Fermi edge disappears abruptly at a critical U value between 4.9 and 5.0 eV for insulating $(V_{0.962}Cr_{0.038})_2O_3$ and between 5.1 and 5.2 eV in the case of metallic V_2O_3. Those

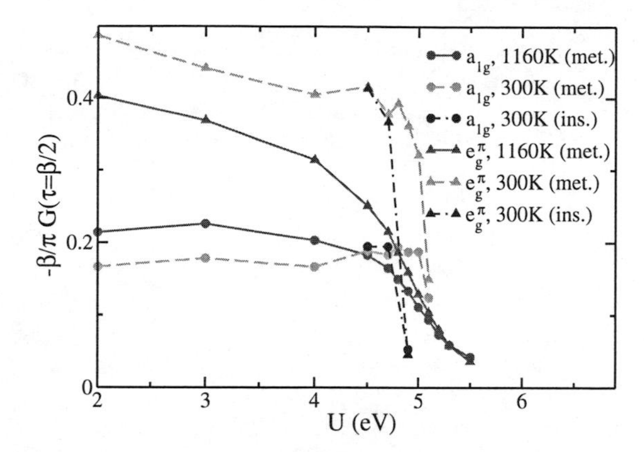

Figure 3.18: Spectral weight of the a_{1g}- and e_g^π-orbitals at the Fermi energy, as estimated by $-\beta/\pi G(\tau = \beta/2)$, vs. U.

critical U values agree quite well with the position of the sharp decrease in the e_g^π quasiparticle weight in Fig. 3.17. They also reaffirm our choice of the value $U = 5$ eV for the simultaneous description of both the metallic (V_2O_3) and insulating ($(V_{0.962}Cr_{0.038})_2O_3$) phase. Although in Fig. 3.18, the weight for the e_g^π-bands is about double the weight for the a_{1g}-band at small U values, they are nearly identical in the vicinity of the transition, indicating a transition at the same critical U value for both types of bands as discussed by Liebsch (2004). Since the statistical error we calculated for the 300 K data is smaller than the size of the symbols used in the figure (for the 1160 K data, it is comparable to the linewidth), we omitted error bars in Fig. 3.18. From our analysis, it is clear that both a_{1g}- and e_g^π-bands undergo a simultaneous transition, although the a_{1g} quasiparticle weight remains constant and the effective mass does therefore not diverge. To resolve this puzzle, it is helpful to analyze the DMFT Green function:

$$G_m(\omega) = \int d\epsilon \frac{N_m^0(\epsilon)}{\omega + \mu - \Sigma_m(\omega) - \epsilon}. \tag{3.2}$$

If we assume a Fermi liquid-like self-energy

$$\Sigma_m(\omega) = \text{Re } \Sigma_m(0) + \left.\frac{\partial \text{Re } \Sigma_m(\omega)}{\partial \omega}\right|_{\omega=0} \omega + \mathcal{O}(\omega^2) \tag{3.3}$$

and insert the quasiparticle weight $Z = (1 - \partial \text{Re } \Sigma_m(\omega)/\partial\omega|_{\omega=0})^{-1}$ we can write the Green function as follows:

$$G_m(\omega) = \int d\epsilon \frac{Z N_m^0(\epsilon)}{\omega + Z(\mu - \text{Re } \Sigma_m(0) - \epsilon)} + \mathcal{O}(\omega^2). \tag{3.4}$$

A metal-insulator transition due to electronic correlations can therefore occur in two different ways: if the effective mass diverges ($Z \to 0$, left part of Fig. 3.19) or if the effective chemical potential $\mu - \text{Re}\ \Sigma_m(0)$ moves outside the non-interacting LDA DOS ($N_m^0(\mu - \text{Re}\ \Sigma_m(0)) \to 0$, middle and right part of Fig. 3.19).

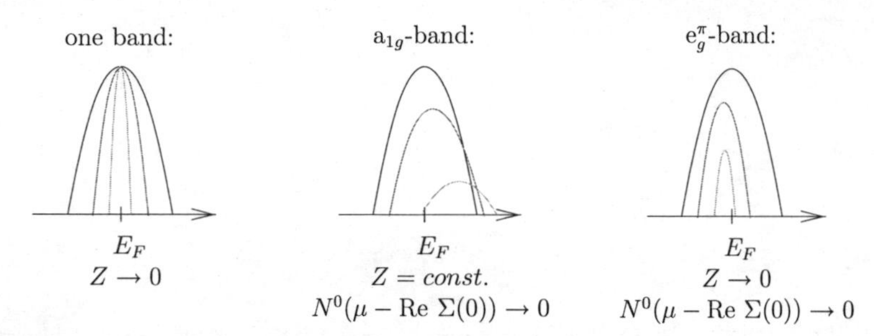

Figure 3.19: Illustration of the different MITs for one band (left) and the a_{1g}-band (middle) and e_g^π-bands (right) of V_2O_3 (for increasing interaction, the color changes from dark to light grey), from Held (2005).

That the latter happens in V_2O_3 is illustrated in Fig. 3.20, where the effective chemical potential is plotted as a function of the Coulomb interaction (where we approximate $\text{Re}\ \Sigma_m(0)$ by its value at the lowest Matsubara frequency $\text{Re}\ \Sigma_m(\omega_0)$).

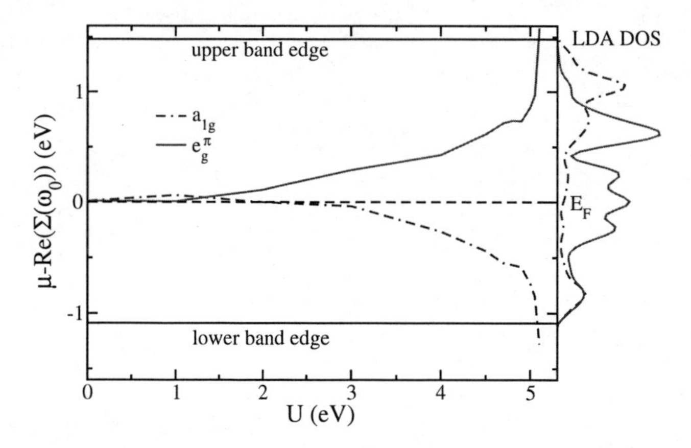

Figure 3.20: Effective chemical potential $\mu - \text{Re}\ \Sigma(\omega_0)$ vs. U. The upper and lower band edges of the non-interacting LDA DOS are shown as solid lines and the entire LDA DOS of V_2O_3 is plotted vertically at the right z-axis.

At the MIT ($U \approx 5.1-5.2$ eV for the LDA DOS for V_2O_3), the effective chemical potential crosses the upper LDA band edge for the a_{1g}- and the lower band edge for the e_g^π-bands. This has important physical consequences for the MIT: in the a_{1g}-band, the height of the quasiparticle peak goes to zero when $N^0_{a_{1g}}(\mu - \text{Re } \Sigma_{a_{1g}}(0)) \to 0$, the width which is given by Z stays finite (middle part of Fig. 3.19). The quasiparticle DOS $N^0_{a_{1g}}(\mu - \text{Re } \Sigma_{a_{1g}}(0))/Z$ does not diverge which leads to non-divergent physical quantities proportional to this DOS like the linear coefficient of the specific heat and the local susceptibility. In the e_g^π-bands, $N^0_{e_g^\pi}(\mu - \text{Re } \Sigma_{e_g^\pi}(0)) \to 0$ and $Z \to 0$, both the height and width of the quasiparticle peak go to zero simultaneously (right part of Fig. 3.19). This peculiar behavior explains the pronounced changes in the orbital occupation in Fig. 3.16 and how an MIT can occur simultaneously in both types of bands although the a_{1g}-quasiparticle weight does not vanish.

3.5 Comparison with experimental spectra

To compare our LDA+DMFT results with experimental photoemission spectra, all theoretical data in this section was obtained by multiplying with the Fermi function at the temperature of the experiments $T \approx 180$ K and broadening them with a 0.09 eV Gaussian to account for the experimental resolution (Mo, Denlinger, Kim, Park, Allen, Sekiyama, Yamasaki, Kadono, Suga, Saitoh, Muro, Metcalf, Keller, Held, Eyert, Anisimov and Vollhardt 2003). For the comparison with x-ray absorption spectroscopy (XAS) data, the same procedure with an inverse Fermi function at $T = 300$ K and a broadening of 0.2 eV was used (Müller, Urbach, Goering, Weber, Barth, Schuler, Klemm, Horn and denBoer 1997) (see appendix A for details). In order to allow a direct comparison of our theoretical spectra and to avoid introducing a further temperature dependence, we used the Fermi function at the same temperature for all theory curves. All LDA+DMFT spectra presented in this section were calculated with the Coulomb interaction $U = 5$ eV. In the experimental photoemission spectra of Mo et al. (2003) and Schramme (2000), the inelastic Shirley-type background and the O-2p contribution were subtracted. All experimental and theoretical spectra were normalized to the same area, i.e. the same occupation in the vanadium t_{2g}-bands. For the PES, the occupation is two electrons per vanadium site, for the XAS, we have four t_{2g}-electrons (the spectrum is measured in units eV^{-1}).

The LDA+DMFT results for different temperatures (1160 K, 700 K and 300 K) are presented in Fig. 3.21. For comparison with the full (unprocessed) spectra, see Fig. 3.12. All three curves show a broad, nearly temperature independent peak at -1.2 eV which corresponds to the lower Hubbard band. Just below the Fermi energy, a well-defined resonance-like structure develops with decreasing temperature which is the remainder of the quasiparticle peak after the multipli-

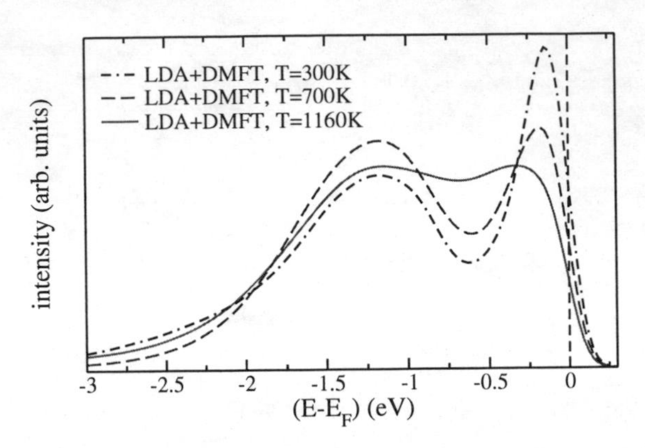

Figure 3.21: LDA+DMFT(QMC) results for the metallic phase at different temperatures at $U = 5$ eV.

cation with the Fermi function. Whereas it is about equal in height to the lower Hubbard band at 1160 K with almost no minimum between the two features, it is 70% higher at 300 K with a pronounced minimum between it and the lower Hubbard band. However, the weight of the residual quasiparticle peak is only increased by 11% from 1160 K to 300 K when measuring from the position of the minimum (-0.63 eV). The main effect of the decreasing temperature is therefore a shift of spectral weight from the region around the minimum at about -0.63 eV to the Fermi edge, resulting in a strong quasiparticle resonance.

In Fig. 3.22, a comparison of our 300 K spectrum with a UPS (ultraviolet photoemission) spectrum by Schramme (2000) and a recent high-resolution bulk-sensitive photoemission spectrum by Mo et al. (2003) is presented. The strong difference between the two experimental spectra is due to the distinctly different photon energies. The data by Schramme (2000) obtained at $h\nu = 60$ eV and $T = 300$ K is very surface sensitive and therefore shows a more insulating behavior, whereas the PES spectrum by Mo et al. (2003) obtained at a much higher photon energy of $h\nu = 700$ eV and $T = 185$ K for the first time exhibits a pronounced quasiparticle peak. Although this is in good qualitative agreement with our 300 K spectrum, the experimental quasiparticle peak has about 11% more spectral weight (0.40 and 0.36 measured from -0.63 eV to the Fermi energy, respectively). The origin of this discrepancy that is especially obvious in the region around the minimum of the theory curve, is presently not clear.

The corresponding calculations for insulating, Cr-doped V_2O_3 together with experimental data by Mo, Kim, Allen, Gweon, Denlinger, Park, Sekiyama, Yamasaki, Suga, Metcalf and Held (2004) are presented in Fig. 3.23. The theoretical

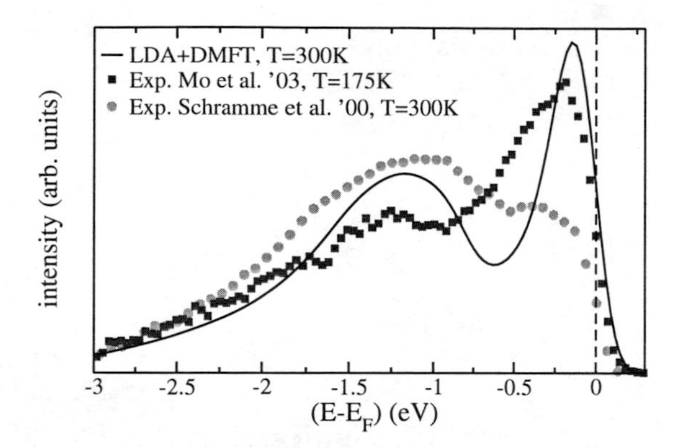

Figure 3.22: Comparison of LDA+DMFT(QMC) results at $T = 300$ K and $U = 5$ eV with photoemission data for metallic V_2O_3 by Schramme (2000) and Mo et al. (2003) with photon energy $h\nu = 60$ eV and $h\nu = 700$ eV, respectively.

spectra show a lower Hubbard band at -1.2 eV similar to the data for the metallic phase, but the quasiparticle peak at the Fermi edge is missing. However, there still remains some spectral weight at the Fermi energy, which is due to highly incoherent states with a large imaginary part of the low-frequency self energy and is not a Fermi liquid effect. This incoherent spectral weight is reduced with decreasing temperatures (as seen in the figure for $T = 1160$ K and $T = 700$ K) and expected to vanish for $T \rightarrow 0$, leading to a increasing resistance with decreasing temperature as expected for an insulator. The experimental data by Mo et al. (2004) for $T = 750$ K and $T = 190$ K is in good qualitative agreement with our theory curves. It shows a reduction of the spectral weight at the Fermi edge with decreasing temperature and a transfer of the spectral weight to the lower Hubbard band. Quantitatively, the theoretical 700 K spectrum has a much sharper lower Hubbard band than the experimental 750 K curve, however, the slope towards the Fermi edge is in good agreement.[9]

Although the comparison with the experimental PES data provides valuable insight into the physics of V_2O_3, it can only give a partial picture, since two thirds of the theoretical spectrum lies above the Fermi energy. The region above E_F is in principle accessible by various methods, e.g. bremsstrahlung isochromat spectroscopy (BIS), inverse photoemission spectroscopy (IPES) or x-ray absorption spectroscopy (XAS). Since IPES data for V_2O_3 is not available and the resolution of BIS is only about 0.6 to 0.7 eV, we compare our theoretical results at

[9] The slope of the Fermi edge is a result of the LDA+DMFT calculation and not of the subsequent multiplication with the Fermi function which only affects the spectrum above the Fermi energy and down to about -0.2 eV.

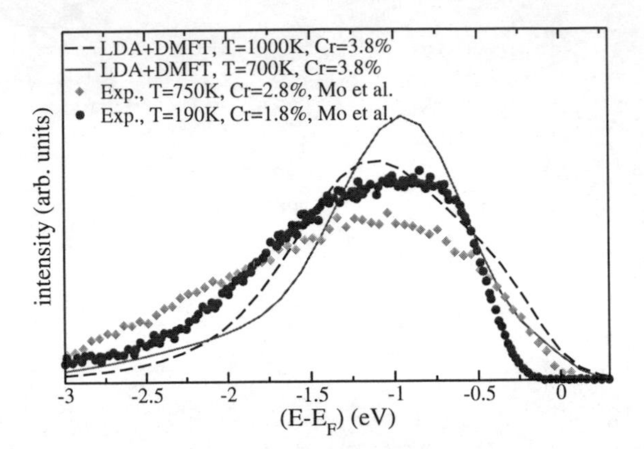

Figure 3.23: Comparison of LDA+DMFT(QMC) results for insulating $(V_{0.962}Cr_{0.038})_2O_3$ at $U = 5$ eV at $T = 700$ K and $T = 1160$ K with photoemission data by Mo et al. (2004) for 1.8% and 2.8% Cr-doping at $T = 190$ K and $T = 750$ K.

$T = 1160$ K, $T = 700$ K and $T = 300$ K with (O-1s) x-ray absorption spectra on the oxygen K-edge for V_2O_3 at 300 K by Müller et al. (1997) (see Fig. 3.24). Since the Fermi edge is not defined in the XAS data, the experimental spectrum was shifted in energy to obtain coinciding peaks at 1.1 eV.[10] All curves were normalized to the same area (corresponding to four t_{2g} electrons). The narrow peak at 1.1 eV and the broad peak at 4.2 eV form the upper Hubbard band which is split due to the Hund's rule coupling. They are almost independent of temperature, but their position and their relative distance depend sensitively on the value of J. Therefore a smaller Hund's rule coupling (e.g. 0.7 eV as shown in Fig. 3.15) would lead to a even better agreement with experiment. Around the Fermi edge, a small shoulder that develops into a peak at low temperatures (300 K) is observed in the theory spectra, this is the residue of the quasiparticle peak. It is puzzling that this feature is missing in the XAS data, since it does not only show up in the theoretical curves but also in the high-resolution PES spectrum by Mo et al. (2003). A shift of the XAS spectrum as proposed by Allen (2004) so that the lower XAS peak would coincide with the quasiparticle peak at the Fermi edge would not resolve this puzzle. To illustrate this, Fig. 3.25 shows the theoretical XAS spectra for Cr-doped V_2O_3 together with the experimental data by Müller et al. (1997) for metallic V_2O_3. The two-peak structure above the Fermi energy and the slope at E_F are in good qualitative agreement in the-

[10] While this procedure is ad hoc, it produces a nearly identical Fermi edge for the insulating theory data and the experimental data in Fig. 3.25.

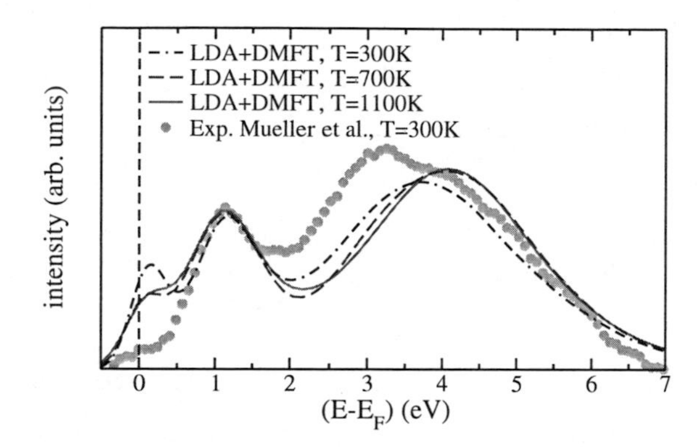

Figure 3.24: Comparison of LDA+DMFT(QMC) results at $U = 5$ eV at $T = 300$ K, $T = 700$ K and $T = 1160$ K with x-ray absorption data at $T = 300$ K by Müller et al. (1997) for metallic V_2O_3.

ory and experiment, although the experiment was done for metallic V_2O_3. This indicates that there is indeed a feature missing in the XAS spectrum and that the metal-insulator transition is not signalled by a shift of the lower XAS peak (as was assumed by Müller et al. (1997)) but by the vanishing of the residue of the quasiparticle peak. Hopefully, additional XAS experiments or, better, inverse photoemission spectroscopy measurements (where the Fermi edge is defined) will shed some light onto this problem.

In Fig. 3.26, the full spectrum below and above the Fermi energy for metallic V_2O_3 for LDA and LDA+DMFT at 300 K is shown together with the experimental PES data by Mo et al. (2004) and the XAS data by Müller et al. (1997). Besides adjusting the value of the Coulomb interaction U so that the correct metallic and insulating behavior is obtained for the respective crystal structures, the LDA+DMFT spectrum was calculated without any fit parameters. Taking this into consideration, the agreement between our theory spectra and the experimental data below and above the Fermi edge is remarkably good. This is especially obvious when comparing to the LDA spectrum, where the same gross features are visible, but their weight, width and position agree neither with LDA+DMFT nor with experiment. It is important to note that the interpretation of the two peaks above the Fermi energy is different in LDA and LDA+DMFT. In LDA, the lower peak stems from the t_{2g}-bands and the upper peak from the e_g^σ-bands as denoted in Fig. 3.26. In LDA+DMFT, the two peaks are a mixture of contributions from the a_{1g}- and the e_g^π-bands with the lower peak mainly due to the a_{1g}- and the upper peak due to the e_g^π-bands. The e_g^σ-bands were not included in our calculations and do not contribute to the correlated spectrum. We

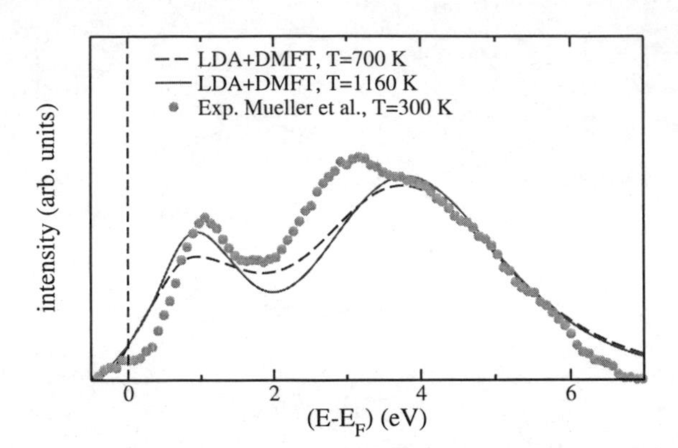

Figure 3.25: Comparison of LDA+DMFT(QMC) spectra for $E > E_F$ for *insulating* $(V_{0.962}Cr_{0.038})_2O_3$ at $U = 5$ eV with x-ray absorption data by Müller et al. (1997) for *metallic* V_2O_3.

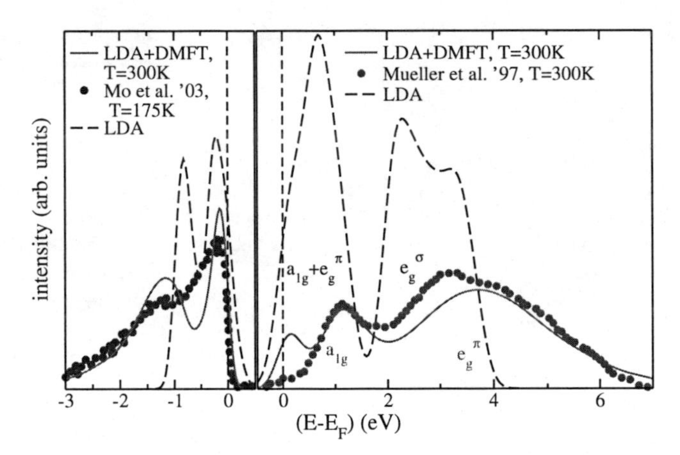

Figure 3.26: Comparison of LDA+DMFT(QMC) results at $U = 5$ eV with PES data by Mo et al. (2003) and x-ray absorption data by Müller et al. (1997) for the metallic phase above and below E_F

therefore have different normalizations of the spectra, the LDA curve is normalized to an area of 10 (corresponding to 6 t_{2g}-electrons and 4 e_g^σ-electrons) whereas the LDA+DMFT data is normalized to an area of 6 (corresponding to six t_{2g}-electrons). In the experiment, a direct assignment of the peaks to certain bands is not possible, hence we used the same normalization as for our LDA+DMFT calculations. The position of the e_g^σ-bands in a correlated calculation can be estimated as follows: Under the assumption that the Coulomb interaction has the same value between t_{2g}-electrons (i.e. U' and $U' - J$ for opposite and aligned spins, respectively) and between t_{2g}- and e_g^σ-electrons and taking into account the distance of roughly 2.5 eV between the centers of gravity of the e_g^σ- and t_{2g}-bands, we expect the e_g^σ-bands (with Hund's rule splitting) at 3.3 eV and 5.1 eV. Therefore, the upper experimental XAS peak will also contain e_g^σ-contributions, i.e. the full upper Hubbard band describes transitions from two-electron configurations $e_g^\pi e_g^\pi$ (with a small admixture of $a_{1g}e_g^\pi$) to three-electron configurations $e_g^\pi e_g^\pi e_g^\pi$, $e_g^\pi e_g^\pi e_g^\sigma$ and $e_g^\pi e_g^\pi a_{1g}$. The larger weight around 3 eV in the experimental compared to the theory curve coincides nicely with the approximate peak position for the lower e_g^σ-band calculated above. Five-band calculations which take into account the e_g^σ-orbitals are in progress. They involve the increased complexity of a calculation with a five-band Hamiltonian (instead of a simple three-band DOS) which seriously increases the computing time necessary for the calculations.

3.6 Conclusion

With the LDA t_{2g}-densities of states for paramagnetic metallic V_2O_3 and paramagnetic insulating $(V_{0.962}Cr_{0.038})_2O_3$ as input, we performed three-band DMFT(QMC) calculations for various U values at 300 K, 700 K and 1160 K. For $U = 5$ eV, the obtained LDA+DMFT spectra are qualitatively different for the two compounds, indicating that a Mott-Hubbard MIT occurs around this value of the Coulomb interaction. Instead of a first-order transition, we only observe a sharp crossover in our calculations at 700 K and 1160 K (at 300 K, the insulating solution could not be calculated with satisfactory accuracy due to computing time restraints). Further, even more extended QMC studies at low temperatures will be necessary to study the first-order MIT and the accompanying discontinuous lattice distortion which leaves the lattice symmetry intact.

Besides the spectral densities, we studied the spin- and orbital state of V_2O_3. We find a $S = 1$ spin state in good agreement with polarization dependent x-ray absorption experiments, but in contradiction to the long-debated $S = 1/2$ model by Castellani et al. (1978). The orbital occupation is predominantly of e_g^π-character, with a small admixture of a_{1g}, in agreement with experiments. For higher values of U (especially at low temperatures), the a_{1g}-occupation decreases in qualitative accordance to the experiment, where a lower a_{1g}-occupation in the insulating phase is observed. Due to the orbital degrees of freedom, the properties of paramagnetic V_2O_3 in the vicinity of the Mott-Hubbard MIT obtained in multi-band LDA+DMFT calculations are remarkably different from those found in the one-band Hubbard model (Georges et al. 1996, Rozenberg et al. 1995, Bulla 1999, Rozenberg et al. 1999). The orbital degrees of freedom are not only responsible for the high asymmetry of the spectra below and above the Fermi energy, they are also required to explain the smallness of the insulating gap and affect the evolution of the quasiparticle peak near the MIT. In the one-band model, the height of the quasiparticle peak is fixed and the quasiparticle weight Z respectively the width goes to zero at the transition. A totally different situation is found in our LDA+DMFT results. For the a_{1g} quasiparticle peak, the width stays constant (indicated by a constant quasiparticle weight Z) but the height goes to zero at the MIT. The e_g^π-bands show a combined reduction of the height and width of Z. Together with the orbital degrees of freedom, the Hund's rule coupling proved to be of crucial importance, especially for the spectrum above the Fermi edge with its split upper Hubbard band and the $S = 1$ spin state. Comparing our theoretical spectra at $U = 5$ eV and $T = 300$ K with PES and XAS measurements, we find good general agreement below and above the Fermi edge. However, the larger weight of the quasiparticle peak in high-resolution bulk sensitive PES and the missing quasiparticle peak in the XAS spectrum remain to be explained and require further theoretical and experimental studies. Further enhancements of the LDA+DMFT scheme (see Sec.1.3), especially the use of the full Hamiltonian (instead of the LDA density of states as in the simplified scheme) and the inclusion of the e_g^σ-orbitals and the oxygen orbitals in the calculations will help to answer the remaining questions in this interesting strongly correlated system.

4. THE SYSTEM $SrVO_3/CaVO_3$

Transition metal oxides (TMO) are at the center of present solid-state research due to the diversity of physical phenomena observed in this class of materials, such as metal-insulator transitions, colossal magnetoresistance and superconductivity. Typically, the 3d-bands are relatively narrow (bandwidth $W \sim 2-3$ eV) in these compounds, which leads to a large ratio of Coulomb interaction to bandwidth U/W. Therefore the electrons are strongly correlated in most TMO, resulting in complicated many-electron physics, which makes realistic calculations difficult. To gain a deeper understanding for the electronic correlations in transition metal oxides, the cubic perovskites with their simple crystal structure and TMO with a $3d^1$-configuration (which do not have a complicated multiplet structure) are especially suited. $SrVO_3$ has both an ideal cubic perovskite structure and a V-$3d^1$-electron configuration and is therefore the perfect starting point for LDA+DMFT studies. Furthermore, Sr can be isovalently substituted by Ca, resulting in a series of compounds $Sr_{(1-x)}Ca_xVO_3$ with $3d^1$-configuration and increasing (orthorhombic) distortion of the crystal structure with rising Ca doping. Among other strongly correlated transition metal oxides, $SrVO_3$ was investigated experimentally early on by Fujimori, Hase, Namatame, Fujishima, Tokura, Eisaki, Uchida, Takegahara and de Groot (1992). In the photoemission spectra (PES) at low photon energies ($h\nu \leq 120$ eV), a pronounced lower Hubbard band was observed which is not found in conventional band structure calculations. Further studies (Aiura, Iga, Nishihara, Ohnuki and Kato 1993, Inoue, Hase, Aiura, Fujimori, Haruyama, Maruyama and Nishihara 1995, Morikawa, Mizokawa, Kobayashi, Fujimori, Eisaki, Uchida, Iga and Nishihara 1995, Inoue, Goto, Makino, Hussey and Ishikawa 1998) investigated the spectral, thermodynamic and transport properties in the $Sr_{(1-x)}Ca_xVO_3$ system and yielded contradictory results. Whereas the thermodynamic properties, i.e., the resistivity, the paramagnetic susceptibility and the specific heat coefficient, are nearly independent of doping between $x = 0$ ($SrVO_3$) and $x = 1$ ($CaVO_3$), the low-energy photoemission spectra change considerably with Ca content. The spectroscopic data seemed to imply that $CaVO_3$ is nearly an insulator whereas $SrVO_3$ is a strongly correlated metal, i.e., that $Sr_{(1-x)}Ca_xVO_3$ is on the verge of a Mott-Hubbard metal-insulator transition. This view was supported by reports of a reduction of the V-O-V bonding angle from $\theta = 180°$ in cubic $SrVO_3$ (Rey, Dehaudt, Joubert, Lambert-Andron, Cyrot and Cyrot-Lackmann 1990) to $\theta = 162°$ in orthorhombic $CaVO_3$ (Jung and Nakotte n.d.), which was believed to effect a strong decrease in bandwidth

(Chamberland and Danielson 1971) and therefore a strong increase in the ratio U/W.

The discrepancy in the experimental results was resolved recently by bulk-sensitive PES measurements by Maiti, Sarma, Rozenberg, Inoue, Makino, Goto, Pedio and Cimino (2001) and especially by Sekiyama et al. (2004). The latter employed high-resolution bulk-sensitive photoemission and showed that the sample surface preparation technique has a strong influence on the results and should preferably be done by fracturing. They further found a strong dependence of the quasiparticle peak on the energy of the x-ray beam and that for that reason a large beam energy is necessary to achieve bulk-sensitivity; at the same time a high instrumental resolution is mandatory to resolve structures like the quasiparticle peak (the resolution was about 100 meV for photon energies of up to 900 eV (Sekiyama et al. 2004)). The results achieved in those experimentally improved PES measurements were found to be almost identical for $SrVO_3$ and $CaVO_3$ (Maiti et al. 2001, Sekiyama et al. 2004). They showed that the nearly insulating spectra for $CaVO_3$ obtained before were due to the surface sensitivity of the PES at lower beam energy and achieved consistency of spectroscopic and thermodynamic results in this compound.

Since $Sr_{(1-x)}Ca_xVO_3$ is strongly correlated and cannot be described accurately in conventional band structure theory, model calculations are important and can provide valuable insight into this class of materials. Rozenberg, Inoue, Makino, Iga and Nishihara (1996) studied the half-filled one-band Hubbard model with a model (Bethe) density of states and used the Coulomb interaction as parameter to fit their spectra to experimental PES. Fig. 4.1 (taken from Rozenberg et al. (1996)) shows the comparison of earlier, more surface-sensitive PES data for $SrVO_3$ and $CaVO_3$ by Inoue et al. (1995) (upper figure) with the fitted theoretical spectra. With a bandwidth of $W = 2$ eV and a Coulomb interaction of $U = 4.8$ eV for $SrVO_3$ and $U = 6.4$ eV for $CaVO_3$, they observed a quite good agreement below the Fermi edge. In their comparison with the LDA results (inset of the lower part of Fig. 4.1), they clearly showed that conventional band structure theory fails to describe $Sr_{(1-x)}Ca_xVO_3$ and that correlation effects are very important in these compounds.

On the other hand, their full theoretical spectrum (see inset of upper part of Fig. 4.1) is symmetric around the Fermi edge (as expected for the Hubbard model at half-filling) and can therefore obviously not describe the complete spectrum of $Sr_{(1-x)}Ca_xVO_3$ (which has a $3d^1$-configuration, i.e., a filling of 1/6 in the three t_{2g}-bands). A realistic description of such systems requires a model that takes into account the orbital degrees of freedom of the partially filled, three-fold degenerate t_{2g}-bands (Fujimori et al. 1992). Hence, in this chapter we employ the LDA+DMFT computational scheme to study $SrVO_3$ and $CaVO_3$ and compare with experimental data below and above the Fermi energy. De-

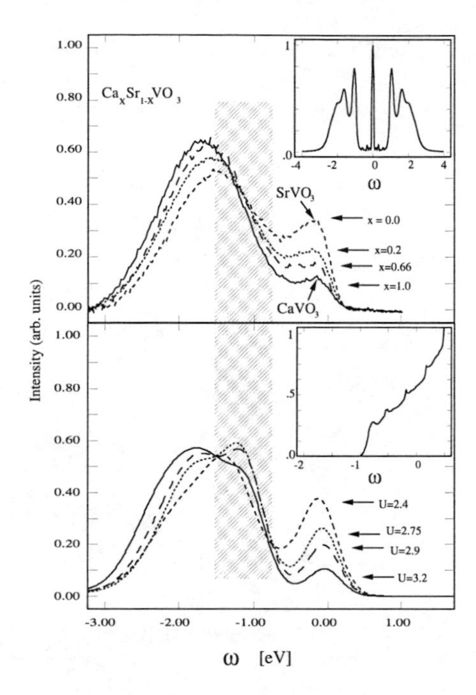

Figure 4.1: Upper figure: Experimental photoemission spectrum of Sr$_{(1-x)}$Ca$_x$VO$_3$ at 80 K from Inoue et al. (1995). Inset: Theoretical spectrum obtained from iterative perturbation theory (IPT) at $T = 0$ K close to the MIT. Lower figure: Theoretical PES spectrum obtained from IPT for various interaction values (U is in units of bandwidth $W = 2$ eV). Inset: Corresponding LDA spectrum from Inoue et al. (1995).

tails of this scheme were discussed in chapter 1. Similar calculations were done independently by Pavarini, Biermann, Poteryaev, Lichtenstein, Georges and Andersen (2004). The effect of enhanced correlations at the surface of SrVO$_3$ due to the reduced coordination number compared to the bulk was studied by Liebsch (Liebsch 2003a, Liebsch 2003b).

4.1 The crystal structures of SrVO$_3$ and CaVO$_3$

SrVO$_3$ is a cubic perovskite with the ideal cubic structure $Pm\bar{3}m$ and lattice constant $a = 3.841$ Å (Rey et al. 1990). The vanadium and strontium atoms form simple cubic subgrids that are shifted against each other in the 111 direction, i.e., each V cube has a Sr atom in the center and vice versa, see Fig. 4.2 (for clarity, only one Sr atom is shown). The oxygen atoms form octahedra around the vanadium atoms in the center. In the ideal structure of $SrVO_3$, those octahedra

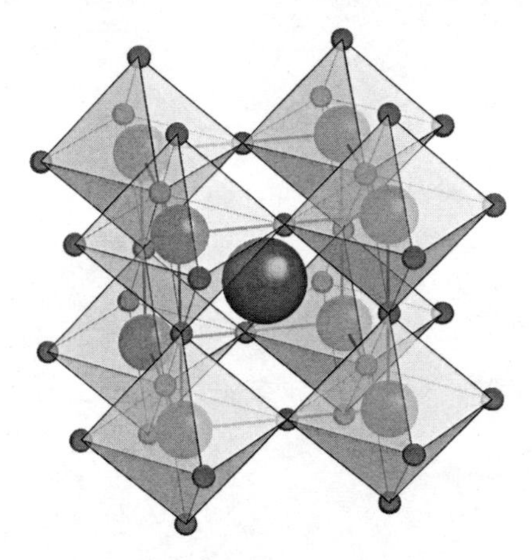

Figure 4.2: Crystal structure of $SrVO_3$. Medium spheres: vanadium; small spheres: oxygen; large sphere: strontium; grey lines with light grey faces: oxygen octahedra; thick black lines: cubic vanadium subgrid; the sizes of the atoms are not true to scale.

are symmetric with equal V-O_{basal} and V-O_{apex} distances and equal O_{basal}-V-O_{apex} and O_{basal}-V-O_{basal} angles of 90°. The VO_6 octahedra are connected via the corners, they are not tilted or rotated against each other. Therefore the angle between two neighboring V atoms via an oxygen atom $\angle$ V-O_{basal}-V ($\angle$ 123 in Fig. 4.3) or $\angle$ V-O_{apex}-V is 180°.

The substitution of Sr^{2+} ions with the isovalent but smaller Ca^{2+} ions leads to a rotation, tilt and distortion of the VO_6 octahedra. $CaVO_3$ therefore has a reduced symmetry with a orthorhombically distorted Pbnm structure (Chamberland and Danielson 1971). In $CaVO_3$, the VO_6 octahedra are rotated by 12° about the c-axis and tilted by 6° from the c-axis in 110 direction (Fig. 4.3). The octahedra itself are also distorted, but the V-O_{basal} and V-O_{apex} distances remain nearly equal.

Due to the rotation and tilt of the oxygen octahedra, the angle between neighboring vanadium atoms $\angle$ V-O_{basal}-V and $\angle$ V-O_{apex}-V is reduced from 180° in $SrVO_3$ to 162° in $CaVO_3$. This decreased bonding angle leads to a reduction in the one-particle bandwidth W and therefore to an increase of the U/W ratio when going from $SrVO_3$ to $CaVO_3$, for details see Sec. 4.2.

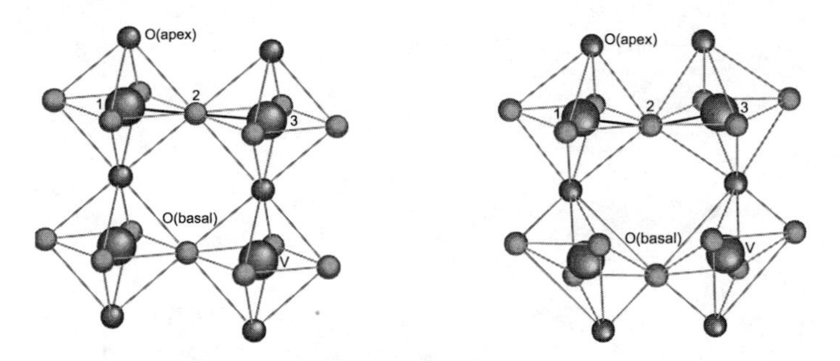

Figure 4.3: Crystal structures of cubic $SrVO_3$ (left) and octahedrally distorted $CaVO_3$ (right). Large spheres: vanadium atoms; small dark spheres: oxygen (apex); small light spheres: oxygen (basal plane). Local c-axis is directed along the V-O_{apex} direction.

4.2 Electronic structure and band structure calculations

In $SrVO_3$ and $CaVO_3$, the V-3d states are situated around the Fermi edge and are partially occupied with one electron per V ion (which has a formal 4+ oxidation) whereas the oxygen 2p-states are fully occupied. In Fig. 4.4 and 4.5, the density of states for both compounds obtained by DFT/LDA calculations based on linearized muffin-tin orbitals (LMTO) (Andersen 1975, Gunnarsson et al. 1983) with an orthogonal basis set are shown.[1] The O-2p states lie between roughly -7.5 eV and -2.0 eV for $SrVO_3$ and for $CaVO_3$, they are separated by a gap of about 1 eV from the V-3d bands which are located between -1.0 eV and 6.5 eV (upper part of Figs. 4.4 and 4.5). The bandwidth is very similar for both compounds, but the shape of the bands and the distribution of the spectral weight is different. Although the main spectral weight of the oxygen and vanadium bands are at different energies, there is also spectral weight of the O-2p bands in the energy region of the V-3d bands and vice versa due to the hybridization between oxygen and vanadium orbitals. The peak at about 1.8 eV, which is especially large in $CaVO_3$ is also due to hybridization, in this case with the Sr-5p and Ca-4p bands, respectively. The vanadium atoms in both $SrVO_3$ and $CaVO_3$ are octahedrally coordinated by oxygen. The five V-3d levels are therefore split into three-fold degenerate t_{2g}-states at lower energy and two-fold degenerate e_g-states at higher energy by the octahedral crystal field, see the diagram in Fig. 4.6. This splitting can also be observed in the LDA results of Figs. 4.4 and 4.5 (lower figures). While for the cubic symmetry of $SrVO_3$ a hybridization of t_{2g} and e_g is forbidden, for the distorted orthorhombic structure of $CaVO_3$ a mixing of those states is pos-

[1] The following orbitals were included in the orbital basis: Sr(5s,5p,4d,4f), Ca(4s,4p,3d), V(4s,4p,3d) and O(2s,2p,3d)

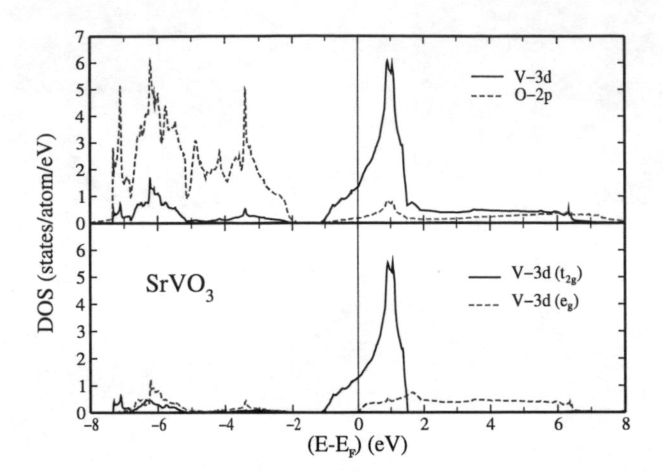

Figure 4.4: Density of states of $SrVO_3$. Upper figure: V-3d and O-2p DOS; Lower figure: partial V-3d (t_{2g}) and V-3d (e_g^σ) DOS.

sible. This can be observed in Fig. 4.5, where the Ca-4p hybridization peak at 1.8 eV is visible for both t_{2g} and e_g. In Fig. 4.4, the respective Sr-5p peak is only seen in the e_g-bands. The t_{2g}-bands of both compounds are located directly around the Fermi edge, between -1 eV and 1.5 eV, neglecting the Sr-5/Ca-4p hybridization peak at 1.8 eV. The e_g-bands extend from the Fermi energy to about 6.5 eV. Although there is a small overlap between t_{2g}- and e_g-bands between 0 and 1.5 eV, the difference of the centers of gravity of the two sets of bands is comparable to the bandwidth, therefore the t_{2g}- and e_g-bands can be considered as well-separated in both compounds for the purpose of our LDA+DMFT calculations. Since the t_{2g}-bands are dominant in the vicinity of the Fermi edge, the low energy physics will be governed by those bands. In the energy range between -8 eV and -2 eV, the aforementioned t_{2g}- and e_g-subbands which are due to the hybridization with the oxygen 2p-states can be seen. They amount to 12% and 15% of all the t_{2g}-states in $SrVO_3$ and $CaVO_3$, respectively.

Fig. 4.7 shows the t_{2g} LDA DOS of $SrVO_3$ and $CaVO_3$ in detail. Despite the appreciable bending of the V-O-V bond angle in $CaVO_3$, we obtain similar LDA spectra for both compounds with only a minor splitting and broadening of the central peak around 1 eV. This also shows in the t_{2g}-bandwidth for $CaVO_3$ ($W_{CaVO_3} = 2.5$ eV), which is only 4% smaller than for $SrVO_3$ ($W_{SrVO_3} = 2.6$ eV). The observed change in bandwidth is consistent with a $|\cos\theta|$ scaling of this parameter ($|\cos 162°| \approx 0.95$) (Harrison 1980). The smaller V-O-V bonding angle has a much stronger effect on the e_g-bands, here the reduction of bandwidth is

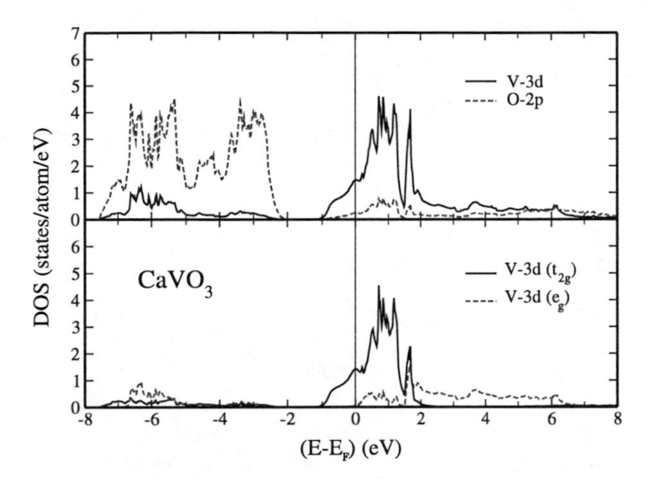

Figure 4.5: Density of states of $CaVO_3$. Upper figure: V-3d and O-2p DOS; Lower figure: partial V-3d (t_{2g}) and V-3d (e_g^σ) DOS.

about 10%. This different behavior of t_{2g}- and e_g-bands can be understood when one calculates the effective t_{2g}-t_{2g} and e_g-e_g hopping parameters. The main contribution to the e_g-e_g hopping is due to a d-p-d hybridization which decreases with the lattice distortion, leading to a smaller hopping and therefore smaller e_g-bandwidth. In the t_{2g}-bands, besides the d-p-d hybridization there is also an important contribution of the direct d-d hybridization. Since the lobes of the t_{2g}-orbitals point more directly towards each other in the distorted structure, the d-d hybridization increases with the distortion. The competition between the decreasing p-d-p and the increasing d-d hybridization leads to a comparably small change in the observed t_{2g}-bandwidth.

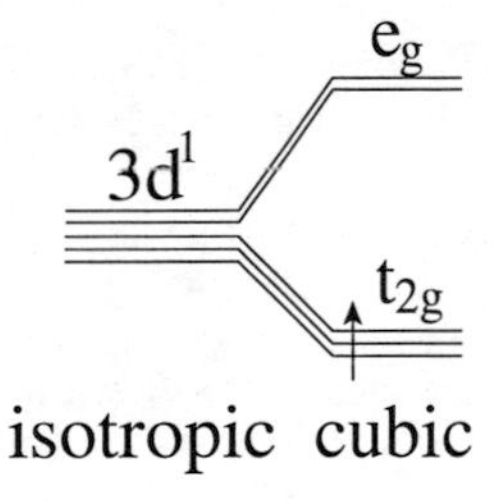

Figure 4.6: Crystal field splitting of the V-3d levels in cubic $SrVO_3$.

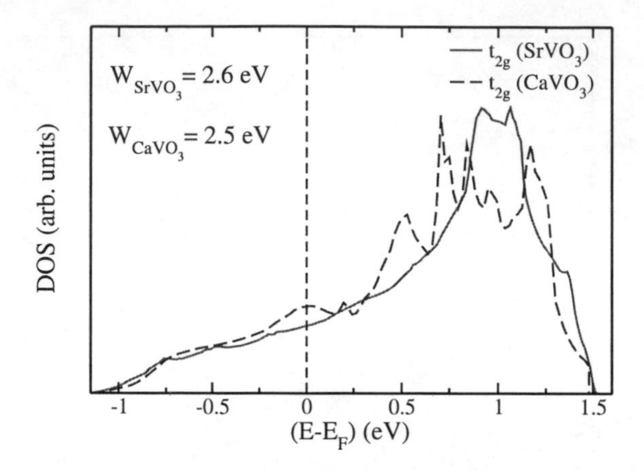

Figure 4.7: t_{2g} density of states of $SrVO_3$ and $CaVO_3$.

Besides the change in bandwidth, the orthorhombic distortion in $CaVO_3$ lifts the degeneracy of the t_{2g}-orbitals, they are split into d_{xy}-, d_{xz}- and d_{yz}-orbitals, see Fig. 4.8. Since this splitting is much smaller than the t_{2g}-bandwidth, it has only a minor effect on the density of states. The e_g-orbitals are more affected by the orthorhombic distortion than the t_{2g}-orbitals as can be seen in the lower part of Figs. 4.4 and 4.5.

In order to perform *ab initio* LDA+DMFT calculations, the values of the Coulomb interaction parameters $\bar{U}$ and J were obtained in constrained LDA calculations (see Sec. 1.1.4) for the t_{2g}-orbitals with the e_g-orbitals participating in the screening (Solovyev et al. 1996). For $SrVO_3$, we obtained $\bar{U} = U' = 3.55$ eV and $J = 1.0$ eV.[2] For $CaVO_3$, the calculation of the Coulomb parameter $\bar{U}$ is problematic because of the mixing of the t_{2g}- and e_g-orbitals which makes it difficult to account for the effect of the e_g screening on $\bar{U}$. Because of its purely atomic character, the Hund's rule coupling J is typically less dependent on screening. Experimental estimates of $\bar{U}$ can be obtained from the optical conductivity where a peak that originates from the transition between the lower and upper Hubbard band can be observed. In optical conductivity measurements in the $Sr_{1-x}Ca_xVO_3$ series of compounds, the position of this peak was found to be independent of x (Makino, Inoue, Rozenberg, Hase, Aiura and Onari 1998). The derived $\bar{U}$ value is in agreement with our result and also with previous calculations in other vanadium systems (Solovyev et al. 1996). Thus we use $\bar{U} = 3.55$ eV and $J = 1.0$ eV for both $SrVO_3$ and $CaVO_3$ in our subsequent LDA+DMFT calculations.

[2] For three orbitals, the average LDA Coulomb interaction $\bar{U}$ is equal to the inter-orbital Coulomb interaction U', see Held et al. (2002) and Zölfl et al. (2000) for details.

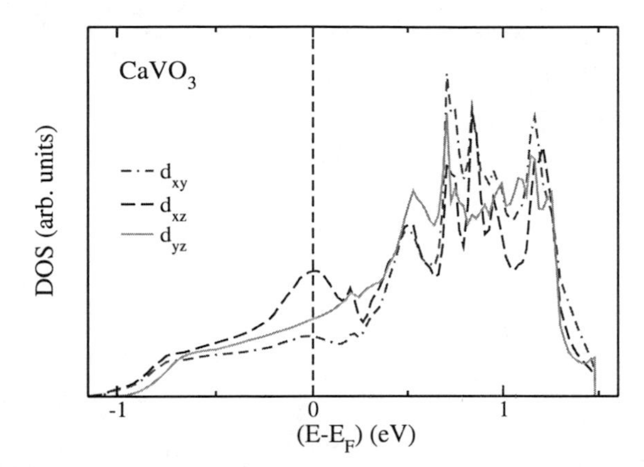

Figure 4.8: t_{2g} density of states of $CaVO_3$ split into d_{xy}, d_{xz} and d_{yz} due to orthorhombic distortion.

4.3 LDA+DMFT spectra

With the non-interacting t_{2g} density of states of $SrVO_3$ and $CaVO_3$ (as shown in Fig. 4.8) as input, we performed three-band DMFT(QMC) calculations at 1160, 700 and 300 K. For the interaction parameters, we used the *ab initio* values $U' = \bar{U} = 3.55$ eV, $J = 1.0$ eV and $U = U' + 2J = 5.55$ eV from constrained LDA calculations.[3] From the resulting imaginary-time QMC Green function, the physically relevant, real-frequency spectral function is obtained with the maximum entropy method, see Sec. 2.3 for details. In Fig. 4.9, the resulting spectra for $SrVO_3$ and $CaVO_3$ at $T = 300$ K are shown. For both materials, the obtained spectra show genuine correlation effects, i.e., a lower Hubbard band at about -2 eV, an upper Hubbard band at roughly 3 eV and a pronounced quasiparticle peak at the Fermi edge. Therefore, $SrVO_3$ and $CaVO_3$ are both strongly correlated metals, they are not on the verge of a Mott-Hubbard metal-insulator transition. Even for unrealistically high values of U,[4] e.g., 6.5 eV for $CaVO_3$, we do not get an insulating gap. Since the quasiparticle peak is nearly vanished at those values of U and those calculations were done at 1160 K to save computing time, we expect an insulating solution for lower temperatures[5] or slightly higher values of U. The spectra of $SrVO_3$ and $CaVO_3$ are very similar. The 4% differ-

[3] The Coulomb repulsion $U = U' - 2J$ is fixed by orbital rotational symmetry.

[4] The values of U from constrained LDA have a typical uncertainty of about 0.5 eV.

[5] With increasing temperature, the Mott-Hubbard gap is filled with spectral weight (Held, Allen, Anisimov, Eyert, Keller, Kim, Mo and Vollhardt 2005), a calculation at lower temperatures would therefore uncover the gap.

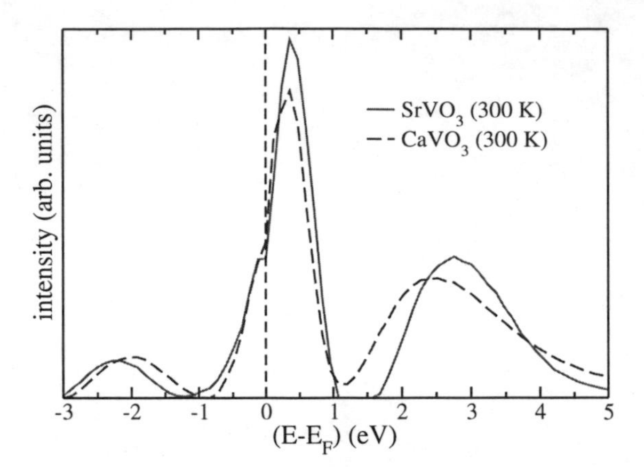

Figure 4.9: LDA+DMFT spectra of $SrVO_3$ and $CaVO_3$ calculated at $T = 300$ K.

ence in LDA-bandwidth is only reflected in slight differences in the position of the Hubbard bands and in an additional transfer of spectral weight from the quasiparticle peak to the Hubbard bands for $CaVO_3$. Therefore, $SrVO_3$ is slightly less correlated than $CaVO_3$ as would have been expected from the LDA DOS or experimental results, but the difference is much smaller than implied from previous photoemission studies (Aiura et al. 1993, Morikawa et al. 1995, Inoue et al. 1995). It is now clear that the insulator-like behavior of $CaVO_3$ observed in those earlier PES experiments can be attributed to surface effects. The question why the surface spectra are so different in $SrVO_3$ and $CaVO_3$ while there bulk spectra are quite similar remains open. Since there is no accurate data for the surfaces of both materials, LDA calculations have not been performed yet. In preliminary investigations of the surfaces of 3d-systems with partly filled t_{2g}-bands, a strong decrease in bandwidth as would be expected was not found. Instead, the results point to the emergence of a quasi one-dimensional surface band, which would be susceptible to correlation effects and could therefore explain the nearly insulating behavior of $CaVO_3$ in surface-sensitive PES experiments (Kozhevnikov n.d.). Using a tight binding approach in connection with DMFT calculations, Liebsch found that although the total bandwidth at the surface is similar to the bulk, the transfer of spectral weight to the center of the band leads to an effective band narrowing at the surface (Liebsch 2003a, Liebsch 2003b).

Since the band-filling is only 1/6 (both compounds have a $3d^1$-electron configuration), the spectra are strongly asymmetric. The center of the quasiparticle peak is about 0.4 eV above the Fermi energy, only 20% of its weight is below the Fermi edge. It is the most prominent feature of the spectrum, 48% of the overall weight is centered there. The lower Hubbard band contributes only 7% to the overall weight, the upper Hubbard band 45%. The different heights of the quasiparticle peaks lead to different effective masses, we get $m^*/m_0 = 2.1$ for $SrVO_3$ and $m^*/m_0 = 2.4$ for $CaVO_3$ which is in good agreement with $m^*/m_0 = 2-3$ for both compounds obtained from de Haas-van Alphen experiments (Inoue, Bergemann, Hase and Julian 2002) and thermodynamics (Aiura et al. 1993, Inoue et al. 1995, Inoue et al. 1998).

The effect of temperature on our LDA+DMFT calculations can be seen in Fig. 4.10, where the spectra for $CaVO_3$ at 1160, 700 and 300 K are shown. With

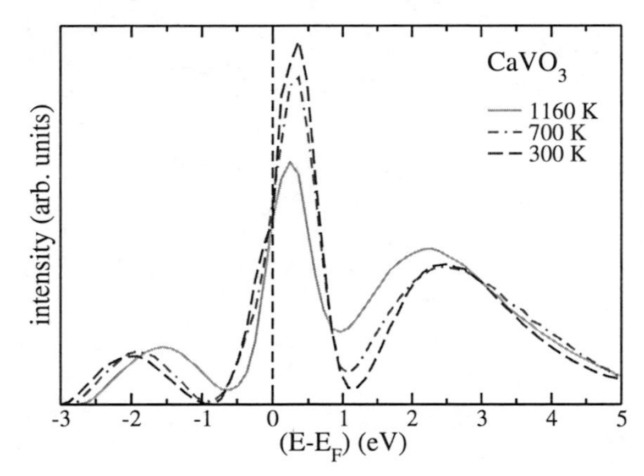

Figure 4.10: LDA+DMFT spectra of $CaVO_3$ at different temperatures.

decreasing temperature, the height of the quasiparticle peak increases markedly, the spectral weight is transferred mainly from the shoulders of the Hubbard bands, especially from the region between 0.8 and 2 eV. The effect of temperature is less pronounced for lower temperatures, i.e., the effect is much stronger between 1160 and 700 K than between 700 and 300 K.

Besides the dependence on temperature, we studied the effect of the Hund's rule coupling. In Fig. 4.11, the $SrVO_3$ spectrum of Fig. 4.9 (with $U = 5.55$ eV and $J = 1.0$ eV) is shown together with a spectrum without Hund's rule coupling ($U = 3.55$ eV and $J = 0$ eV). For both spectra, the inter-orbital coupling is identical, $U' = 3.55$ eV. In contrast to systems like V_2O_3, where the Hund's rule coupling has a major influence on the spectrum (see Sec. 3.3 for details), the spectra for $SrVO_3$ are nearly identical, with only a minor transfer of weight from the shoulders of the quasiparticle peak to the respective Hubbard bands. A splitting of one of the Hubbard bands is not visible since the there is only one electron in the t_{2g}-bands and thus, the splitting is too small (only J instead of $2J$

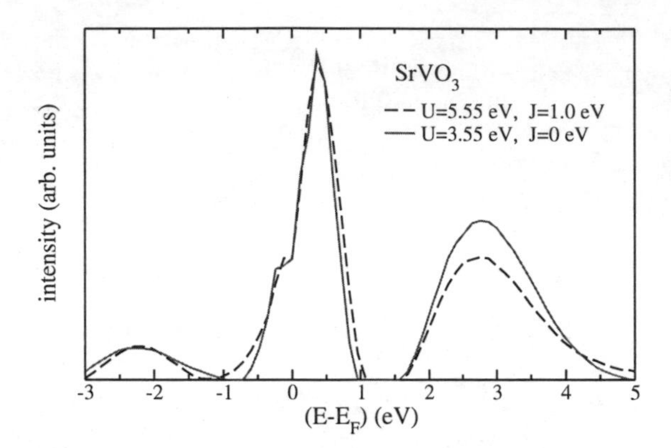

Figure 4.11: LDA+DMFT spectra of $SrVO_3$ at $T = 300$ K and $U' = 3.55$ eV with ($J = 1.0$ eV) and without ($J = 0$ eV) Hund's rule coupling.

as in, e.g., V_2O_3) to lead to two distinct peaks.[6] However, since the Hund's rule coupling is important for other physical quantities like the spin state, it cannot be neglected in the calculations.

4.4 Comparison with experimental spectra

For the comparison of our theoretical data with the experimental spectra from photoemission, we multiplied our LDA+DMFT results with the Fermi function at the experimental temperature ($T = 20$ K) and broadended with a 0.05 eV Gaussian to account for the experimental resolution. The same procedure with an inverse Fermi function at $T = 80$ K and a 0.25 eV Gaussian was applied for a comparison with x-ray absorption spectroscopy (XAS).[7]

Fig. 4.12 shows the comparison of our theoretical curves for $SrVO_3$ and $CaVO_3$ with high-resolution, bulk-sensitive photoemission spectroscopy (PES) data by Sekiyama et al. (2004). The latter were obtained by subtracting estimated oxygen and surface contributions. We find good overall agreement of the theoretical and experimental spectra. Clearly, the considerable orthorhombic distortion caused by the substitution of Sr by Ca has only a minor effect on the spectrum below

[6] A calculation with an artificially enlarged Hund's exchange $J = 2$ eV lead to a splitting of the upper Hubbard band similar to V_2O_3.

[7] We used 0.05 and 0.25 eV instead of the value for the experimental resolution of 0.1 eV (Sekiyama et al. 2004) (PES) and 0.36 eV (Inoue 2004) (XAS), respectively, for the broadening to get coinciding slopes at the Fermi edge, see appendix A for details.

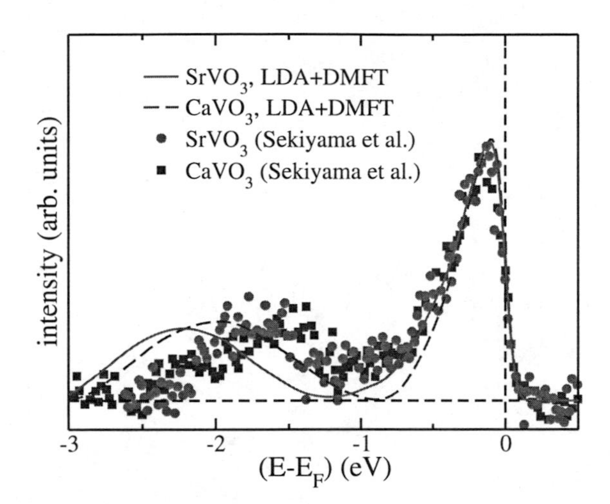

Figure 4.12: Comparison of the LDA+DMFT spectra of $SrVO_3$ and $CaVO_3$ below the Fermi edge with experimental bulk-sensitive high-resolution photoemission spectroscopy (PES) by Sekiyama et al. (2004).

the Fermi edge, i.e., on the PES. The quasiparticle peaks in Fig. 4.12 are nearly identical for $SrVO_3$ and $CaVO_3$ both in theory and experiment. Furthermore, the respective theoretical and experimental peaks agree quite well with only a minimal difference in height. The larger height of the $SrVO_3$ quasiparticle peak of the full theory spectrum has practically no influence on the spectrum below the Fermi energy. In contrast to the good agreement for the quasiparticle peaks, the position of the lower Hubbard bands differs notably, in the theory curves they are about 0.5 eV lower than in experiment. The difference may be partly due to the subtraction of the estimated oxygen contribution which might also remove some spectral weight of the V-3d bands below -2 eV. Furthermore, the uncertainty in the *ab initio* calculated value of $\bar{U}$ is about 0.5 eV, a smaller $\bar{U}$ would shift the lower Hubbard band closer to the Fermi energy. The tendency that the lower Hubbard band of $CaVO_3$ is at a higher energy than of $SrVO_3$ is found both in theory and experiment.

The comparison of our theoretical spectra (for $SrVO_3$ and $CaVO_3$) above the Fermi edge with XAS data by Inoue, Hase, Aiura, Fujimori, Morikawa, Mizokawa, Haruyama, Maruyama and Nishihara (1994) (for $SrVO_3$ and $Ca_{0.9}Sr_{0.1}VO_3$) is shown in Fig. 4.13. Here, the discrepancies between theory and experiment are distinctly larger than for the PES data. The position of the quasiparticle peaks is in good agreement, and the tendency for a smaller quasiparticle peak for the

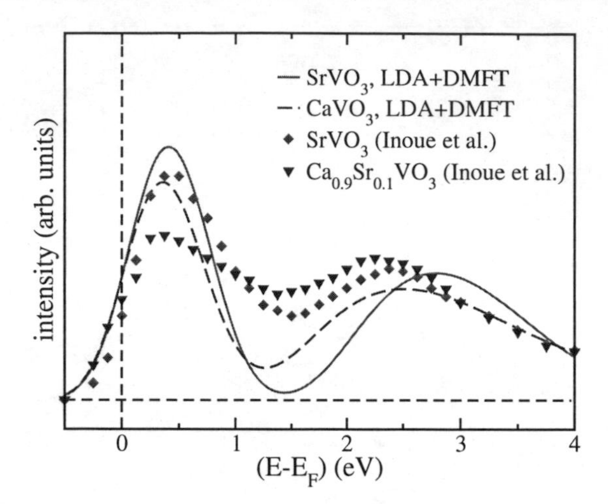

Figure 4.13: Comparison of the LDA+DMFT spectra of $SrVO_3$ and $Ca_{0.9}Sr_{0.1}VO_3$ above the Fermi edge with experimental x-ray absorption (XAS) measurements by Inoue et al. (1994).

more correlated $CaVO_3$ is reproduced correctly. Furthermore, the height of the upper Hubbard bands is comparable. Their position in theory is at slightly higher energy which is consistent with the PES comparison and again indicates that a slightly smaller value of $\bar{U}$ could lead to even better agreement. However, the theoretical curves have a pronounced minimum between the quasiparticle peak and the upper Hubbard band (with nearly zero intensity in the case of $SrVO_3$). In the experimental curves this minimum is only about 30% lower in intensity than the upper Hubbard bands. Compared to the theory, spectral weight is transferred from the quasiparticle peaks into the region of this minimum, leading to a reduced height of those peaks in experiment.

Although there are some differences between theory and experiment, those have to be expected since both the LDA+DMFT and the subtraction procedure to remove oxygen and surface contributions to get a bulk spectrum involve approximations. Furthermore, no parameters were used to fit the theory curves to the experimental data. Hence, LDA+DMFT is able to provide quite accurate *ab initio* spectra both below and above the Fermi edge for $SrVO_3$ and $CaVO_3$. This is a considerable improvement over simple one-band models, where the spectrum can only be fitted and two different sets of fit parameters are used for comparison with PES and XAS data, respectively.

4.5 Conclusion

Starting with the non-interacting density of states for $SrVO_3$ and $CaVO_3$ from LDA and the Coulomb interaction parameters $\bar{U} = 3.55$ eV and $J = 1.0$ eV from constrained LDA calculations, we obtained correlated spectra for various temperatures (300 K, 700 K and 1160 K) from three-band LDA+DMFT(QMC) calculations without any fit parameters. Although the V-O-V bonding angle is considerably reduced when going from the cubic $SrVO_3$ to the distorted $CaVO_3$ structure, the respective LDA bandwidth is only decreased by 4%. Accordingly, the LDA+DMFT spectra are also very similar. Both systems are strongly correlated metals with a strong quasiparticle peak at the Fermi edge and clearly separated lower and upper Hubbard bands. The difference in the quasiparticle peaks is also apparent in the different effective masses for both compounds, which are in agreement with experimental values from de Haas-van Alphen experiments (Inoue et al. 2002) and thermodynamics (Aiura et al. 1993, Inoue et al. 1995, Inoue et al. 1998). The temperature dependence is small for the incoherent part of the spectrum, only the height of the incoherent peak at the Fermi energy increases with lower temperature. The Hund's rule coupling J is too small in $SrVO_3$ and $CaVO_3$ to lead to a visible multiplet structure in these $3d^1$-systems. $CaVO_3$ is slightly more correlated than $SrVO_3$, leading to a smaller quasiparticle peak at the Fermi energy. The theoretical PES spectra are very similar since the differences of the quasiparticle peaks are visible only above the Fermi edge and agree well with high-resolution bulk-sensitive PES experiments (Sekiyama et al. 2004). The agreement is less good for the comparison with XAS spectra (Inoue et al. 1994), but the main features and the differences between the two compounds are reproduced there as well. Overall, the results of our LDA+DMFT calculations describe the experimental results well both below and above the Fermi energy without any fitting parameters and are therefore a considerable improvement over simple one-band model approaches and pure LDA calculations.

5. LiV_2O_4 - A 3D HEAVY-FERMION SYSTEM

In the last years, LiV_2O_4 was the topic of numerous experimental and theoretical studies. The interest in this system was initiated by the discovery of heavy-fermion behavior with a spin fluctuation temperature $T_K \sim 28$ K by Kondo, Johnston, Swenson, Borsa, Mahajan, Miller, Gu, Goldman, Maple, Gajewski, Freeman, Dilley, Dickey, Merrin, Kojima, Luke, Uemura, Chmaissen and Jorgensen (1997). LiV_2O_4 was the first d-electron system to exhibit heavy-fermion behavior. Typically, heavy-fermion materials are intermetallic compounds of heavy elements with f-electrons like cerium or uranium and have strongly renormalized effective masses $m^* \approx 100 - 1000\, m_e$ (Andres, Graebner and Ott 1975, Stewart 1984) and a strongly enhanced spin susceptibility χ at low temperatures. For LiV_2O_4, Kondo et al. (1997) reported a large electronic specific heat coefficient $\gamma \approx 0.42$ J/(mol K^2) which is much larger than in other metallic transition metal compounds (e.g., for $V_{2-y}O_3$, $\gamma \lesssim 0.047$ J/(mol K^2) (Carter, Rosenbaum, Metcalf, Honig and Spalek 1993)). With decreasing temperature, the system changes from local-moment behavior to renormalized Fermi-liquid behavior. In the temperature region between 50 K and 1000 K, the experimental magnetic susceptibility fits well to a Curie-Weiss law with small negative Curie temperature, i.e., a weak antiferromagnetic spin interaction (Kondo et al. 1997, Urano, Nohara, Kondo, Sakai, Takagi, Shiraki and Okubo 2000). Down to 0.02 K, no magnetic ordering was observed. The electrical resistivity ρ has a strong T^2 dependence $\rho = \rho_0 + AT^2$ with an enormous A that scales with γ^2 as in conventional heavy-fermion systems (Urano et al. 2000).

LiV_2O_4 has been studied theoretically in band structure calculations by various groups with different LDA (local density approximation) implementations, among others by Eyert, Höck, Horn, Loidl and Riseborough (1999) with the scalar-relativistic augmented spherical wave basis (ASW), Anisimov, Korotin, Zölfl, Pruschke, Le Hur and Rice (1999) in standard LMTO basis, and with the LDA+U method and both Matsuno, Fujimori and Mattheiss (1999) and Singh, Blaha, Schwarz and Mazin (1999) with a full potential, scalar-relativistic implementation of the linear augmented plane wave (LAPW) method. Besides those *ab initio* studies, various model calculations on the system were performed. Burdin, Grempel and Georges (2002) studied the competition between the Kondo effect and frustration of exchange interactions in a Kondo lattice model within DMFT and a large-N approach for the spin liquid. Kusunose, Yotsuhashi and

Miyake (2000) investigated a two-band Hubbard model in slave-boson mean-field approximation. In a simple two-band model (Hopkinson and Coleman 2002b) and a t-J model with strong Hund's rule coupling for the d-electrons (Hopkinson and Coleman 2002a), Hopkinson *et al.* observed a two-stage screening in LiV_2O_4. A comparison of LiV_2O_4 and $LiTi_2O_4$ with typical heavy-fermion systems was done by Varma (1999). Johnston (1999) gave an overview over the various experimental and theoretical research on LiV_2O_4. Considerable efforts have been made to explain the heavy-fermion behavior of LiV_2O_4 at low temperatures, but a undisputed picture has not yet evolved. One explanation attempt is based on a mechanism similar to the one used for systems with local moments on a pyrochlore lattice which are frustrated with respect to the antiferromagnetic interactions. Another explanation by Anisimov et al. (1999), which does not involve frustration, is based on the idea that the electrons in the partially filled t_{2g}-bands separate into localized ones forming local moments and delocalized ones forming a metallic, partially filled band. The hybridization between those two types of electrons are thought to be the source of the heavy-fermion behavior, similar to the conventional heavy-fermion compounds with their localized f-electrons. This idea, which was studied by Anisimov et al. (1999) within LDA+U, will be reevaluated on the basis of our LDA+DMFT calculations in this chapter.

5.1 The crystal and electronic structure of LiV_2O_4

LiV_2O_4 has a fcc normal-spinel structure with space group Fd3m (face centered, point group C_{3v}); no structural phase transitions were found down to 4 K. It has a (cubic) lattice constant of 8.2361 Å at 200 K which is decreasing monotonically on going to lower temperatures (Chmaissem, Jorgensen, Kondo and Johnston 1997). In Fig. 5.1, the crystal structure of LiV_2O_4 is shown. Its fcc unit cell contains two LiV_2O_4 formula units with four V atoms, two Li atoms and eight O atoms. The oxygen atoms form slightly distorted edge-shared octahedra with an oxygen position parameter of $x_0 = 0.26111$ (compared to the ideal value of $x_0 = 0.25$) with vanadium atoms in the center. The edge-sharing V octahedra form layers of parallel chains, successive layers are rotated by 90° and are interconnected through the top of the vanadium octahedra. In the cavities between the chain layers the lithium atoms are located, they are tetrahedrally coordinated by oxygen atoms. Lithium and vanadium atoms form a LiV_2 substructure identical to the AB_2 structure of the cubic Laves phase (C15) which shows a strong frustration of the local moments on the B sites (Wada, Shiga and Nakamura 1989). Although LiV_2O_4 has cubic space group, the local point group symmetry of the vanadium atoms in the structure is trigonal (D_{3d}). The four V atoms in the unit cell form a tetrahedron with the trigonal axes of the vanadium directed towards its center (Fig.5.2).

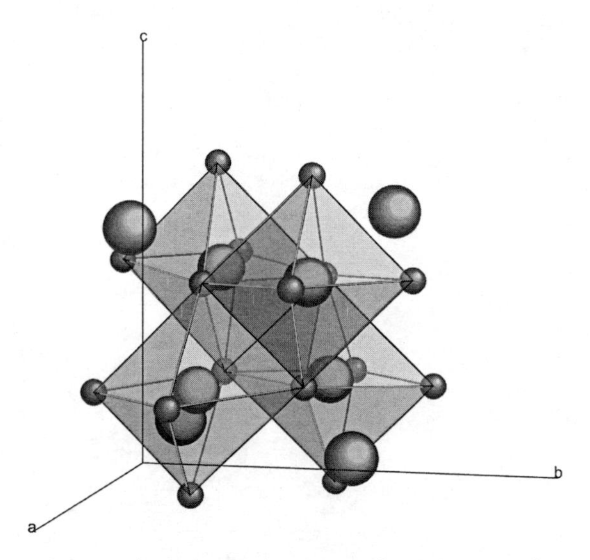

Figure 5.1: The normal-spinel crystal structure of LiV_2O_4; V: medium spheres; O: small spheres; Li: large spheres; grey lines denote the oxygen octahedra.

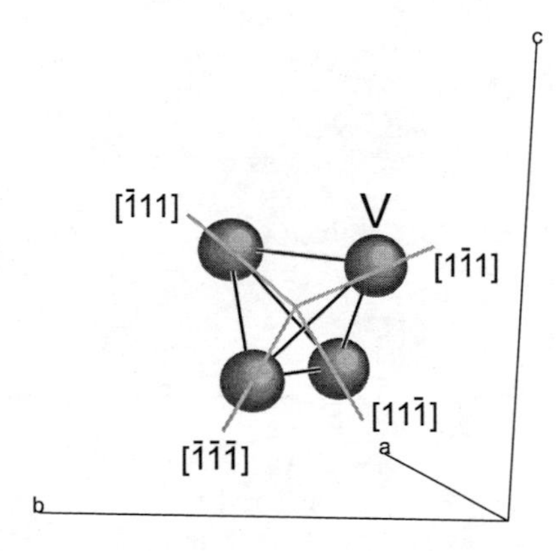

Figure 5.2: Vanadium tetrahedron formed in the spinel unit cell with the trigonal axes for each V atom.

In a octahedral crystal field, the five V-3d bands are split into three degenerate t_{2g}- and two degenerate e_g^σ-bands[1] (Fig. 5.3). The Fermi level lies within the t_{2g}-bands which are therefore partially filled, while the e_g^σ-bands are empty and the lower lying O-2p bands are fully occupied. Since the formal valence of the V atoms in LiV_2O_4 is 3.5+, there are six valence electrons per unit cell or 1.5 per vanadium, leading to a non-integer occupation of the t_{2g}-bands and a metallic state as observed experimentally by Reuter and Jaskowsky (1960). Due to the

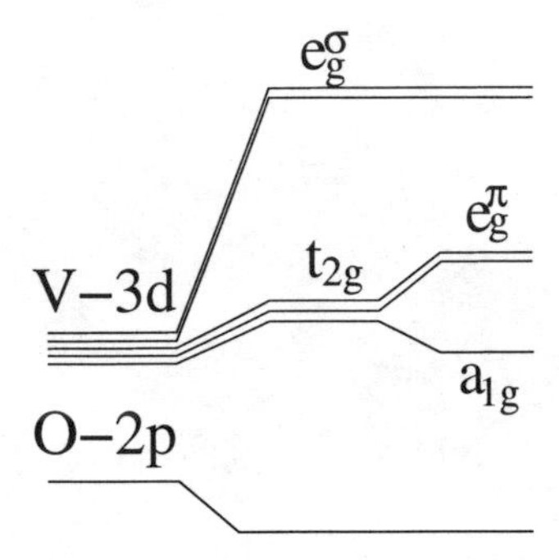

Figure 5.3: Schematic electronic level splitting in the trigonally distorted octahedral crystal field.

trigonal distortion of the octahedral field in LiV_2O_4, the t_{2g}-levels are split into one a_{1g}- and two degenerate e_g^π-bands (similar to V_2O_3). Since this splitting is small compared to the bandwidth of the respective bands, the t_{2g}-bands are still overlapping. Matsuno et al. (1999) reported a splitting of ~ 0.14 eV compared to an overall t_{2g}-bandwidth of ~ 2.2 eV.

5.2 Results from LDA calculations

In a first step to get more information on the electronic structure of LiV_2O_4, first principle DFT/LDA calculations based on linear muffin-tin orbitals (LMTO) were performed by Nekrasov and Pchelkina (2002). More information on density functional theory can be found in Sec. 1.1. For the LDA calculations, the TB-LMTO-ASA program by Jepsen and Andersen (2000) was used; the muffin-tin radii were 2.00 a.u. for Li, 2.05 a.u. for V and 1.67 a.u. for O. In order to improve the accuracy of the calculations, the overlap between atomic spheres was set to zero and the number of empty spheres was increased accordingly. In Fig. 5.4, the

[1] The designations t_{2g}, e_g and a_{1g} originate from group theory (see Hammermesh (1989) for details).

obtained partial LDA density of states of LiV_2O_4 is shown in the energy region around the Fermi edge. In the upper spectrum, three well-separated groups of bands are discernible. The fully occupied O-2p bands are in the energy range from -8 eV to -3 eV, the partially filled V-3d t_{2g}-bands are between -1.0 eV and 0.8 eV and the completely empty V-3d e_g^σ-bands between 2.3 eV and 3.2 eV. Although

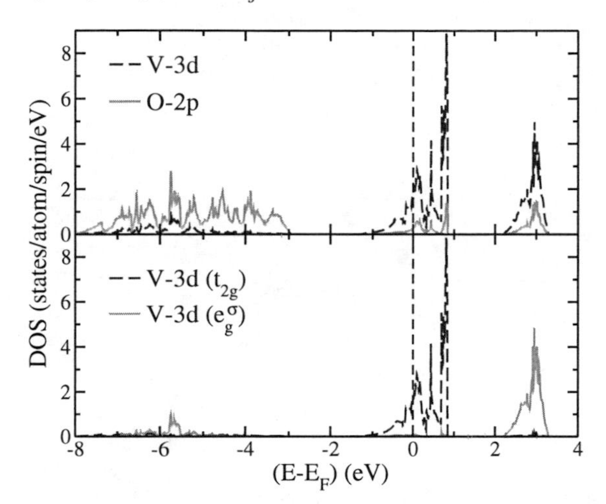

Figure 5.4: LMTO DOS of LiV_2O_4. Upper figure: O-2p and V-3d contributions; lower figure: partial V-3d t_{2g} and e_g^σ densities of states.

the p-d hybridization between the oxygen and the vanadium states is apparent from the V-3d contributions to the O-2p bands and the admixture of the oxygen states in the vanadium bands, it is small compared to other early transition metal oxides (e.g., V_2O_5 (Eyert and Höck 1998)). The main contribution of the V-3d states to the oxygen bands comes from the unoccupied e_g^σ-states which form σ-bonds with the O-2p states and strongly overlap with them. The t_{2g}-orbitals, which form π-bonds with the oxygen only yield a negligible admixture to the O-2p bands. Besides this p-d bonding, the t_{2g}-orbitals also have strong σ-bonds to the t_{2g}-orbitals of the neighboring vanadium sites in the fcc sublattice. Due to those different types of bonding, V-V σ-bonds and V-O π-bonds for the t_{2g}- and V-O σ-bonds for the e_g^σ-orbitals, the bandwidths of those two groups of bands are also different (2.05 eV for t_{2g} and 0.9 eV for e_g^σ). The t_{2g}- and e_g^σ-bands are completely separated with practically no hybridization between them (see the lower part of Fig. 5.4).

In Fig. 5.5, the partial densities of states of the a_{1g}- and e_g^π-bands are shown. The two types of t_{2g}-bands are distinctly different: the bandwidth of the a_{1g}-

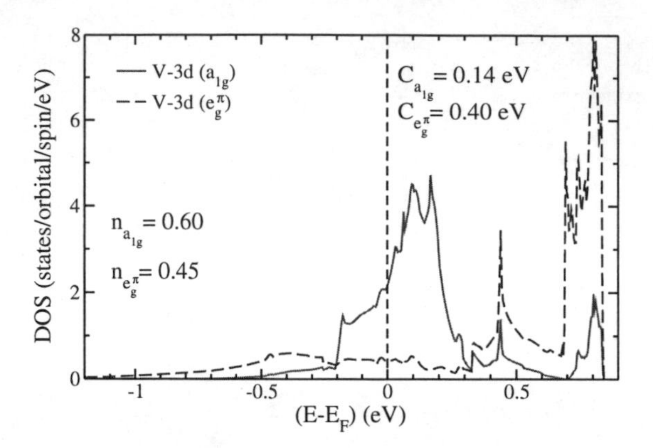

Figure 5.5: Partial a_{1g} and e_g^π DOS of LiV_2O_4 with the filling n and the centers of gravity C of the respective bands.

band ($W_{a_{1g}}$ = 1.35 eV) is 50% smaller than that of the e_g^π-band ($W_{e_g^\pi}$ = 2.05 eV). Whereas the a_{1g}-band is concentrated mainly around the Fermi level between -0.2 and 0.3 eV, the e_g^π-bands exhibit a long, flat tail from the lower band edge up to 0.3 eV with the main spectral weight from 0.3 eV to the upper band edge. This is also reflected in the center of gravity of the bands ($C_{a_{1g}}$ = 0.14 eV, $C_{e_g^\pi}$ = 0.40 eV), it is distinctly lower for the a_{1g}-band. Since the contribution of the tail in the e_g^π-bands is small, the occupation of the a_{1g}-band $n_{a_{1g}}$ = 0.6 is slightly higher than that of the e_g^π-bands $n_{e_g^\pi}$ = 0.45. Eyert et al. (1999) performed LDA calculations (with the oxygen position parameter x_0 = 0.26111 of LiV_2O_4 and with the ideal value of x_0 = 0.25 for the spinel structure) and found that the a_{1g}-band is mainly participating in weak π-type V-O-bonding and is very sensitive to changes in the oxygen position, i.e., to the value of the trigonal splitting. Although the trigonal splitting is considerably smaller than the bandwidth, it is important for the understanding of the strongly correlated system LiV_2O_4. When strong electronic correlations (with an interaction parameter larger than the bandwidth as found by Anisimov et al. (1999) in LDA+U calculations) are taken into account, it depends on the value and sign of the trigonal splitting which t_{2g}-orbital is localized. Since the trigonal splitting is very sensitive to the accuracy of the calculations, the LDA computations were performed without overlap between atomic spheres and with additional empty spheres. If one defines the trigonal splitting as the energy difference between the centers of gravity of the a_{1g}- and e_g^π-bands, one finds a splitting of 0.26 eV and can conclude that the lower-lying a_{1g}-band is more favorable for localization when strong correlations are taken into account.

A good measure for the electronic correlations of a system is the effective mass m^*. It can be obtained from the low-temperature electronic specific heat coefficient γ. The ratio of the effective mass m^* to the band mass m_b can be derived from the experimental specific heat coefficient γ and the LDA-calculated specific heat coefficient $\gamma^{LDA}=\pi^2 k_B^2 N_A D(E_F)/3$ via $m^*/m_b=\gamma/\gamma^{LDA}$. With the experimental value taken from Kondo et al. (1997) and the LDA value calculated with the LDA DOS $D(E_F)$ at the Fermi energy, we find $m^*/m_b \approx 25.8$, in good agreement with previous results (Matsuno et al. 1999). The huge renormalization of the quasiparticle mass m^* underlines the necessity for taking into account the electronic correlations to correctly describe the physics of LiV_2O_4 and justifies the designation of lithium vanadate as a heavy-fermion system.

5.3 Results from correlated calculations

For the LDA+DMFT calculations, we used the DMFT code described in Sec. 1.2.2. The density of states of the a_{1g}- and e_g^π-orbitals presented in Fig. 5.5 was used as input for the DMFT. Furthermore, we took the averaged d-d interaction parameter $\bar{U} = 3.0$ eV and the Hund's rule coupling parameter $J = 0.8$ eV from constrained LDA calculations (see Sec. 1.1.4) by Anisimov et al. (1999). In the rotational invariant case this translates to an intra-orbital coupling of $U = 4.6$ eV and an inter-orbital coupling of $U' = 3.0$ eV ($U = U' - 2J$). The total number of electrons in the t_{2g}-bands was set to 1.5, only the t_{2g}-orbitals were taken into account in the calculations. Within our QMC calculations, the temperature was approximately 750 K which strikes a good balance between computing time and comparability with experimental results (as the calculations for V_2O_3 and $SrVO_3$ for $T = 1160$ K to $T = 300$ K showed, the temperature dependence decreases rapidly when going to lower temperatures). In Fig. 5.6, the spectral functions for the a_{1g}- and e_g^π-orbitals derived from the QMC data by analytical continuation with maximum entropy (see Sec. 2.3) are presented. The spectrum shows the typical correlation effects with a splitting of the spectrum and a formation of a lower Hubbard band (at -3 eV for the a_{1g}- and -1.5 eV for the e_g^π-bands) and an upper Hubbard band (at 3 eV for the a_{1g}- and 4 eV for the e_g^π-bands). The slight differences between the e_g^π-bands are due to the statistical fluctuations in the QMC and the analytical continuation with the maximum entropy method. The strong asymmetry between the a_{1g}- and the e_g^π-bands in the LDA spectra is retained in the DMFT results and is not reduced by the correlations. This is even more apparent when we directly compare the LDA and DMFT results for the respective orbitals as shown in Fig. 5.7. Both types of bands show a considerable transfer of spectral weight from the energy region around the Fermi edge to the lower and especially to the upper Hubbard bands. The spectra are considerably broadened due to correlations (note that the DMFT spectra are magnified in

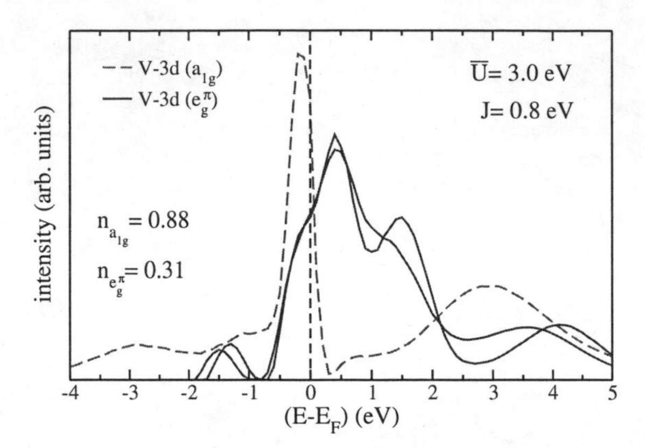

Figure 5.6: LDA+DMFT spectra for the a_{1g}- and e_g^π-bands of LiV_2O_4.

intensity by a factor of 5 in Fig. 5.7). The centers of gravity are shifted to higher energies. Nevertheless, the central peak of the a_{1g} spectrum is slightly displaced to lower energies in the correlated calculations. This leads to a considerable shift in the electron occupations, with $n_{a_{1g}}$ increasing from 0.6 to about 0.9 and $n_{e_g^\pi}$ decreasing from 0.45 to 0.3 per band. Taking into account the correlations via DMFT leads to a nearly localized electron in the a_{1g}-orbital and to two e_g^π-orbitals close to quarter-filling. Similar results were found by Anisimov et al. (1999). In their LDA+U calculations (which only take the correlations into account via a static Hartree term), they found $n_{a_{1g}} = 0.84$ by using an artificial antiferromagnetic order, whereas our DMFT calculations are done in the physically correct paramagnetic phase and are therefore represent a substantial improvement over the former work.

In contrast to single-band DMFT spectra, the DOS in Fig. 5.6 shows a complicated peak structure. In order to understand the origin of those features, it is helpful to examine the spectrum of the atomic Hamiltonian and the energy of the one-particle excitations. Since we have 1.5 electrons in the three t_{2g}-bands and U and J are large, the ground state will mainly be a mixture of two states, one with one electron in the a_{1g}-band ($|\sigma\rangle|0\rangle|0\rangle$ with an energy of $\epsilon_{a_{1g}}$) and the other with one electron in the a_{1g}-band and one spin-aligned electron in one of the two e_g^π-bands ($|\sigma\rangle|\sigma\rangle|0\rangle$, $|\sigma\rangle|0\rangle|\sigma\rangle$ with an energy of $2\epsilon_{a_{1g}}+\Delta\epsilon+U'-J$, where $U' = U-2J$ due to rotational symmetry and $\Delta\epsilon \approx 0.26$ eV is the trigonal splitting between a_{1g}- and e_g^π-orbitals). Since those states should be nearly degenerate, we can calculate the energy $\epsilon_{a_{1g}} = -U' + J - \Delta\epsilon \approx -2.5$ eV. From this ground-state energy, one can easily construct the possible single particle excitations, which are shown in Table 5.1.

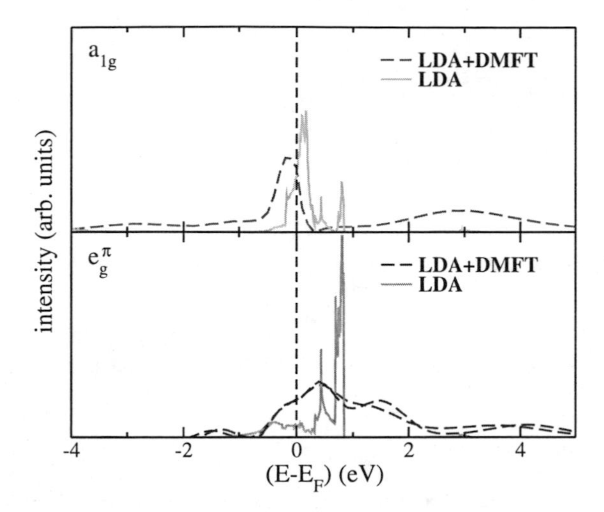

Figure 5.7: Comparison of LDA and LDA+DMFT spectra for the a_{1g}-bands (upper figure) and e_g^π-bands (lower figure); LDA+DMFT results are magnified by a factor of 5.

For the a_{1g}-orbital, there are two excitations with $\omega < 0$, one from a singly occupied to an unoccupied state and a second from a doubly occupied state (which is built from a mixture of a_{1g}- and e_g^π-states, i.e., a mixed valence state) to a singly occupied orbital. Since the excitation energy of the latter process is given by the trigonal splitting $\Delta\epsilon$, its value can be directly calculated form experimental photoemission spectra.

The excitation processes listed in Table 5.1 can be directly mapped to the spectral features in Fig. 5.7, but due to the high temperature for which the QMC calculations were performed, the particular excitations are not easily discernible in the LDA+DMFT spectrum. It is therefore instructive to compare with the results from resolvent perturbation theory and NCA (see, e.g., Zölfl et al. (2000)) which allows not only a distinction between the a_{1g}- and e_g^π-orbitals but also a direct identification of the different initial particle numbers of the excitations. In Fig. 5.8, the single particle excitations for singly, doubly and triply occupied initial states from a NCA calculation are plotted. The energies of the excitations of Table 5.1 are marked by dotted lines and are clearly corresponding to the dominant peaks in the figure. All the peaks for the a_{1g}-orbital have corresponding peaks in the e_g^π-orbitals shifted to higher energies by the trigonal splitting $\Delta\epsilon$, but the weight of those corresponding peaks is in some cases extremely different (e.g., for the peak at $\omega = J$) due to the different band fillings for the respective orbitals.

a_{1g}-orbital	
Excitation process	Excitation energy ω
$\lvert\sigma\rangle\lvert 0\rangle\lvert 0\rangle \rightarrow \lvert 0\rangle\lvert 0\rangle\lvert 0\rangle$	$\epsilon_{a_{1g}}$
$\lvert\sigma\rangle\lvert\sigma\rangle\lvert 0\rangle \rightarrow \lvert 0\rangle\lvert\sigma\rangle\lvert 0\rangle$	$\epsilon_{a_{1g}} + U' - J = -\Delta\epsilon$
$\lvert\sigma\rangle\lvert 0\rangle\lvert 0\rangle \rightarrow \lvert 2\rangle\lvert 0\rangle\lvert 0\rangle$	$\epsilon_{a_{1g}} + U$
e_g^π-orbitals	
Excitation process	Excitation energy ω
$\lvert\sigma\rangle\lvert 0\rangle\lvert 0\rangle \rightarrow \lvert\sigma\rangle\lvert\sigma\rangle\lvert 0\rangle$	$-\epsilon_{a_{1g}} - \Delta\epsilon - U' + J = 0$
$\lvert\sigma\rangle\lvert\sigma\rangle\lvert 0\rangle \rightarrow \lvert\sigma\rangle\lvert 0\rangle\lvert 0\rangle$	$\epsilon_{a_{1g}} + \Delta\epsilon + U' - J = 0$
$\lvert\bar{\sigma}\rangle\lvert 0\rangle\lvert 0\rangle \rightarrow \lvert\bar{\sigma}\rangle\lvert\sigma\rangle\lvert 0\rangle$	$\epsilon_{a_{1g}} + \Delta\epsilon + U' = J$

Tab. 5.1: Single-particle excitations for the atomic model, for spin-degenerate processes only one configuration is listed.

Comparing the LDA+DMFT result (Fig. 5.7) with the NCA plot (Fig. 5.8), a

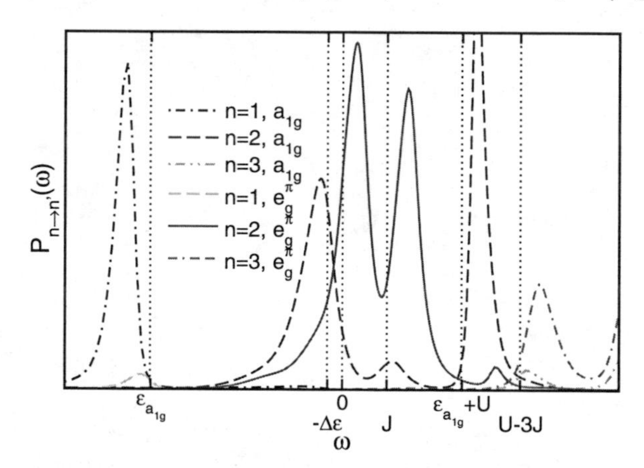

Figure 5.8: NCA density of states separated in singly, doubly and triply occupied states for the a_{1g}- and e_g^π-orbitals, from Pruschke (2004).

close correspondence of the main features in both figures is found. Therefore, the peaks can simply be explained by the atomic multiplet structure. The shifts in energy are renormalizations which typically occur in quantum impurity models. It is interesting to note that the upper Hubbard band of the a_{1g}-orbital is mainly due to a configuration with two electrons whereas in the e_g^π-orbital, it is due to the excitation into a triply occupied state with an energy of $U - 3J$ ($\lvert\sigma\rangle\lvert\sigma\rangle\lvert 0\rangle \rightarrow \lvert\sigma\rangle\lvert\sigma\rangle\lvert\sigma\rangle$). The double peak structure of the e_g^π-orbital between 0 and 2 eV and of the a_{1g}-orbital between -1 and 1 eV is due to two processes with different spin alignment, i.e., the Hund's rule splitting.

5.4 Magnetic properties and comparison with experiments

Since in LDA+DMFT, we have nearly one electron localized in the a_{1g}-band, the local moment corresponds to $S = 1/2$ per V atom and therefore a Curie-like susceptibility. Due to the remaining 0.5 electrons per vanadium atom in the two e_g^π-bands, there will be a small temperature independent Pauli contribution to the susceptibility. This is in agreement with experiments (Kondo et al. 1997, Kondo, Johnston and Miller 1999, Urano et al. 2000), where a paramagnetic Curie-Weiss susceptibility $\chi(T) = C/(T - \Theta) + \chi_0$ with a T-independent part χ_0 (containing core diamagnetic, Pauli paramagnetic and orbital Van Vleck contributions) gives the best fit to the data for temperatures $T = 50 - 1000$ K.

To make closer contact to experiments, we performed magnetic susceptibility calculations within LDA+DMFT(QMC) for a set of temperatures with $\beta = 5, 7, 9, 10, 11, 12, 13$ eV^{-1} ($\approx 2300 - 900$ K) and different magnetic fields $B = 0.005, 0.01, 0.02$ eV. The results[2] are shown in Fig. 5.9 for a Hund's rule coupling of $J = 0$ eV (diamonds) and $J = 0.8$ eV (triangles) together with the experimental data and a fit to a Curie-Weiss law[3]

$$\chi(T) = N_A \frac{\mu_B^2 p_{eff}^2}{3k_B(T - \Theta)}. \tag{5.1}$$

Extrapolation yields a Curie-Weiss temperature of $\Theta \approx 640$ K for $J = 0.8$ eV and $\Theta \approx -450$ K for $J = 0$ eV, the effective paramagnetic moment $p_{eff} = \sqrt{g^2 S(S+1)}$ calculated from the fitted Curie-Weiss constant is 1.65 and 1.3, respectively. This reduced value of 1.65 for the effective moment compared to a spin-1/2 system (which has $p_{eff} = 1.73$) can be attributed to the a_{1g}-occupation which is only 0.88 for $J = 0.8$ eV and thus leads to a spin $S < 1/2$. With the smaller occupation taken into account, the ideal system would yield $p_{eff} = 1.52$, which is smaller than the LDA+DMFT value because the the e_g-orbitals are not taken into account. For $J = 0$ eV, the a_{1g}- and e_g-orbitals do not couple via the local exchange interaction. With the reduced a_{1g}-occupation of 0.75 in this case, the LDA+DMFT value of 1.3 agrees with the value for the spin-1/2 system ($p_{eff} = 1.73 \times 0.75 = 1.3$). Comparing our $J = 0.8$ eV calculations to experiment,

[2] For high temperatures, the dominant contribution to the experimental susceptibility is of Curie-Weiss type, we therefore neglected the Pauli contribution in our analysis of the high-temperature LDA+DMFT calculations.

[3] In Byczuk and Vollhardt (2002), the Curie-Weiss law was analytically derived in DMFT for the Hubbard model.

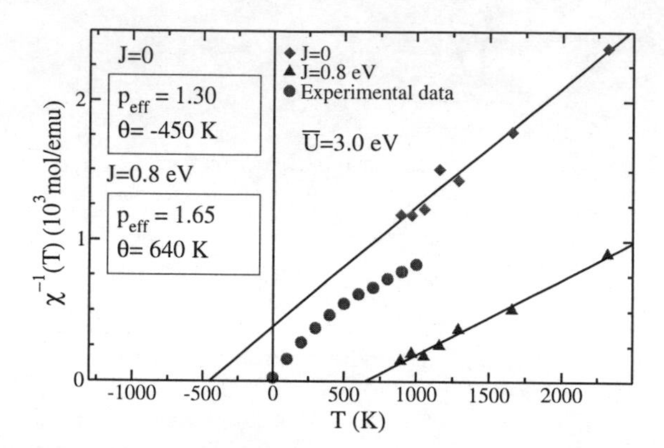

Figure 5.9: Inverse spin susceptibility χ^{-1}(T) obtained by LDA+DMFT calculations at $J = 0$ and $J = 0.8$ eV.

we find good agreement for the effective moment but a large ferromagnetic Curie-Weiss temperature of 640 K, in contrast to the small antiferromagnetic Curie-Weiss temperature in the range from -20 to -60 K from experiment (Kondo et al. 1997, Kondo et al. 1999, Urano et al. 2000). Similarly to our results, Anisimov et al. (1999) obtained a ferromagnetic intersite exchange parameter $J_{dex} = 530$ K, which is the sum of direct and double exchanges. This ferromagnetic exchange coupling in LiV_2O_4 can be easily understood in the double exchange picture. A necessary precondition for double exchange, localized electrons forming local moments and delocalized ones in a partially filled, broad band, leading to a strong ferromagnetic coupling between the local moments, is fulfilled in LiV_2O_4 with one of its electrons nearly localized in the a_{1g}-band and a filling of $1/8$ in each of the broad e_g^π-bands.

Another important factor for the double exchange is the Hund's rule coupling which energetically favors spin alignment, i.e., ferromagnetic order. Without this intra-atomic exchange, the double exchange coupling vanishes. This can be seen in the $J = 0$ calculations of Fig. 5.9, where we obtain a negative Curie-Weiss temperature, i.e., an effective antiferromagnetic exchange coupling. Obviously, there is a subtle competition between ferromagnetic double exchange from the hybridization of the e_g^π-orbitals and antiferromagnetic direct exchange resulting from a_{1g}-a_{1g} hybridization in LiV_2O_4. In our DMFT calculations with $J = 0.8$ eV, the ferromagnetic contribution is dominating. Even slightly smaller J values as found in high-energy spectroscopy experiments ($J \approx 0.65$ eV)[4] do not significantly

[4] Zaanen and Sawatzky (1990) derived a value of $J = 0.64$ eV from Racah parameters for the V ions, Mizokawa and Fujimori (1996) found $J = 0.68$ eV in their cluster-model analysis of photoemission spectra.

change our results.

However, since we only use the LDA DOS and not the (downfolded) Hamiltonian as input to our DMFT calculations, we completely neglect the a_{1g}-e_g^π hybridization. Anisimov et al. (1999) observed an antiferromagnetic Kondo coupling between the a_{1g}- and e_g^π-orbitals due to this hybridization with a coupling $J = -630$ K. Therefore there are three competing exchange interactions present between the a_{1g}- and e_g^π-electrons in LiV_2O_4: ferromagnetic double exchange ($J \approx 1090$ K), antiferromagnetic direct exchange ($J \approx -450$ K) and antiferromagnetic Kondo exchange ($J \approx -630$ K, not taken into account in our calculations). These three contributions effectively cancel and lead to a small Curie-Weiss temperature around 0 K as observed in experiment. The competition between the different exchange contributions can also help to explain neutron scattering experiments, where a switch from antiferromagnetic to ferromagnetic spin fluctuations is observed at $T \approx 40$ K (Krimmel, Loidl, Klemm, Horn and Schober 1999, Krimmel, Loidl, Klemm, Horn and Schober 2000a, Krimmel, Loidl, Klemm, Horn and Schober 2000b, Fujiwara, Yasuoka and Ueda 1998, Mahajan, Sala, Lee, Borsa, Kondo and Johnston 2000). This temperature coincides with the energy scale of the Kondo coherence temperature. Whereas below this temperature, the antiferromagnetic Kondo and direct exchanges prevail, above T_{coh} the Kondo effect diminishes and the ferromagnetic double exchange dominates. Another indication for this competition from experiment is the slightly enhanced g-factor, which points towards ferromagnetic coupling between the conduction electrons and the local moments (Johnston 1999, Johnston, Ami, Borsa, Crawford, Fernandez-Baca, Kim, Harlow, Mahajan, Miller, Subramanian, Torgeson and Wang 1995), in contrast to the negative Curie temperature which indicates antiferromagnetic interactions.

5.5 Conclusion

Within LDA and LDA+DMFT, we investigated the electronic structure, orbital state and magnetic properties of LiV_2O_4. While the small trigonal splitting between the a_{1g}- and e_g^π-bands in the non-interacting density of states obtained by LDA does not lead to a separation of the bands, it produces a significant difference in the centers of mass and the effective bandwidths of the orbitals. In the subsequent LDA+DMFT calculations, we found that of the 1.5 3d-electrons per V atom, one is nearly localized in the a_{1g}-orbital whereas the two e_g^π-orbitals form relatively broad, metallic bands with a filling of 1/8.

The temperature dependence of the calculated paramagnetic susceptibility is in accordance to the Curie-Weiss law derived from experiments, the obtained effective magnetic moment of 1.65 is also in good agreement. The small negative Curie temperature in experiment can be explained by the cancellation of three competing contributions to the effective exchange interaction in the V-3d orbitals, the ferromagnetic double exchange, the antiferromagnetic direct exchange and the antiferromagnetic Kondo exchange. Since the last contribution is due to a coupling between a_{1g}- and e_g^π-bands which was not taken into account in our present calculations, we found a large ferromagnetic Curie temperature. In future calculations with an input LDA Hamiltonian which includes the a_{1g}-e_g^π hybridization, the three competing contributions to the exchange may likely almost cancel, leading to a small residual antiferromagnetic Curie-Weiss temperature in accordance to experiment.

SUMMARY

The LDA+DMFT method is one of the most promising approaches for an parameter-free *ab initio* description of strongly correlated electronic systems at present. In this computational technique, the band structure obtained from a local density approximation (LDA) calculation and the Coulomb interaction parameters obtained from a constrained LDA calculation are used as input to a multi-band Hubbard model calculation within dynamical mean-field theory (DMFT). The complicated multi-band Anderson impurity problem that arises in the DMFT self-consistency cycle is solved by numerically exact quantum Monte Carlo simulations. With this recently developed approach, the spectra and other physical properties like the orbital occupation or the quasiparticle weight were calculated for the correlated transition metal oxide systems $(V_{1-x}Cr_x)_2O_3$, $Sr_{(1-x)}Ca_xVO_3$, and LiV_2O_4. These systems all have a V-3d shell that is split by the crystal field of the octahedrally coordinated oxygen into three partially filled t_{2g}-orbitals and two empty e_g^σ-orbitals. The filling of the t_{2g}-orbitals and the splitting between the two groups of orbitals as well as between the t_{2g}-orbitals is different and thus leads to distinctly different physical properties. The LDA+DMFT calculations were performed for the three t_{2g}-bands.

For V_2O_3 and $(V_{0.962}Cr_{0.038})_2O_3$, a sharp crossover between the metallic and the insulating phase was found for intermediate Coulomb interaction (U = 5 eV). Due to the high temperatures of the QMC calculations ($T = 700, 1160$ K for the computations in the insulating phase and additionally $T = 300$ K for the metallic phase), the first-order metal-insulator transition could not be investigated. The obtained multi-band LDA+DMFT spectra show a strong asymmetry with a splitting of the upper Hubbard band due to the Hund's rule coupling. Thus, a part of the upper Hubbard band is shifted towards the Fermi edge, which is the cause for the smallness of the insulating gap in spite of a large Coulomb interaction. The orbital degrees of freedom and the Hund's rule coupling also change the character of the metal-insulator transition compared to the one-band Hubbard model. Whereas for the one-band case the quasiparticle weight goes to zero at the transition with a fixed quasiparticle height, in the multi-band case only the width goes to zero for the a_{1g}-band or both the width and height go to zero for the e_g^π-bands. The investigation of the spin and orbital state of vanadium oxide revealed a $S = 1$ spin state and a predominant occupation of the e_g^π-bands. Those results are in accordance to experimental results but in contrast to former theoretical results, especially to the $S = 1/2$ picture by Castellani et al. (1978) which predicts a complete occupation of the a_{1g}-band. The comparison of the theoretical results with experimental photoemission and x-ray absorption spectra yielded good general agreement. The discrepancies that arise for the quasiparticle peak have to be reexamined in extended theoretical and experimental studies and will hopefully yield new insights into this classic Mott-Hubbard system.

With the Coulomb interaction parameters from constrained LDA and the density of states from LDA calculations, LDA+DMFT spectra for $SrVO_3$ and $CaVO_3$ were obtained at $T = 300, 700$ and 1160 K without any fit parameters. Both materials are strongly correlated metals far from a metal-insulator transition. $CaVO_3$ is found to be slightly more correlated with a smaller quasiparticle peak in the spectrum and also with a smaller quasiparticle weight. Consequently, the effective masses for both systems are also slightly different, in agreement with experimental results. A splitting of the upper Hubbard band due to the Hund's rule coupling as in V_2O_3 could not be observed in $SrVO_3$ and $CaVO_3$, since in these d^1-systems, the energy difference is not sufficient to yield two separate peaks. The theoretical PES spectra are very similar and in good agreement with high-resolution photoemission spectra. In the comparison with x-ray absorption spectra, the differences are more pronounced, but the general features and the differences between both materials are captured by the theoretical results.

The third transition metal oxide compound that was studied in this work is the heavy-fermion system LiV_2O_4. Due to the trigonal splitting between the a_{1g}- and e_g^π-bands, the centers of mass of the LDA densities of states are significantly different which leads to a nearly localized electron in the a_{1g}-band and two broad, metallic e_g^π-bands with a filling of 1/8. The LDA+DMFT results for the paramagnetic susceptibility yielded a temperature dependence in accordance to a Curie-Weiss law and an effective moment in agreement with experimental data. The large ferromagnetic Curie temperature obtained in the calculations is in contrast to the small negative (antiferromagnetic) Curie temperature in experiment. The reason for this discrepancy is that the antiferromagnetic Kondo exchange between the nearly localized a_{1g}-band and the itinerant e_g^π-bands is missing in the LDA+DMFT. Future calculations within the Wannier function formalism, which include the hybridization between the a_{1g}- and e_g^π-orbitals and thus contain all three exchange contributions, may be able to address this problem.

The LDA+DMFT(QMC) has proven to be reliable approach for the realistic investigation of strongly correlated multi-band systems. Further advances and refinements of the method will broaden the scope of its applicability and help to give new insights into the physics of this interesting, intensively studied class of materials.

APPENDIX

A. CALCULATION OF THEORETICAL PES AND XAS SPECTRA

To illustrate the calculation of photoemission (PES) and x-ray absorption (XAS) spectra from the full theoretical spectrum, the procedure is shown for the example of a typical theory curve. It was obtained from a three-band DMFT calculation with a Bethe DOS ($U = 5.0$, $J = 0.93$ eV) and a subsequent maximum entropy calculation (Fig. A.1). To obtain a PES curve, the spectral function $A(\omega)$ ob-

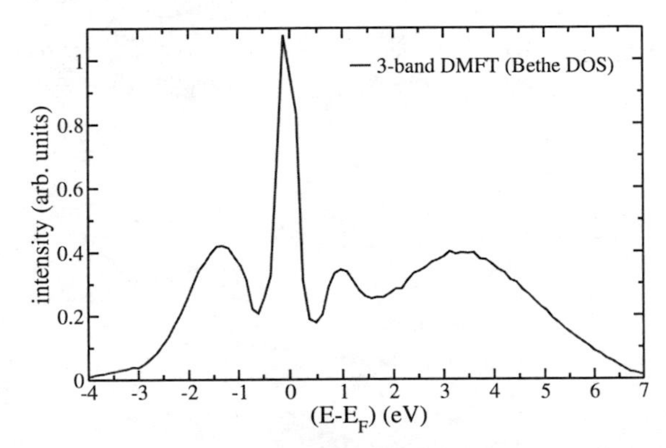

Figure A.1: Full theory spectrum obtained with Bethe DOS and $U = 5.0$, $J = 0.93$ eV.

tained from maximum entropy is first multiplied with the Fermi function $f_T(\omega)$ at temperature T:

$$A_{PES}(\omega) = A(\omega) f_T(\omega) = A(\omega) \frac{1}{1 + e^{\frac{\omega - \mu}{k_B T}}} \tag{A.1}$$

The dashed curve in the left part of Fig. A.2 shows the result of this calculation for $T = 300$ K. For the XAS data, the inverse Fermi function, which has a minus sign in the exponent, has to be used (dashed curve in right part of Fig. A.2).

Afterwards, the curves are broadened with a Gaussian with a standard deviation

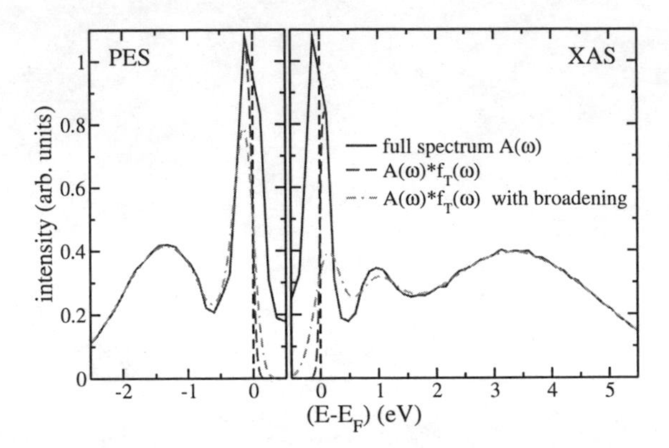

Figure A.2: Full theory spectra (full lines), spectra multiplied with the Fermi function at 300 K (dashed lines) and spectra multiplied with the Fermi function and broadened with a Gaussian with standard deviation 0.1 eV (PES, left side) and 0.2 eV (XAS, right side) to account for the experimental resolution (dash-dotted curves).

of D:[1]

$$A'_{PES}(\omega) = \int d\omega' \frac{e^{\frac{(\omega-\omega')^2}{2D^2}}}{\sqrt{2\pi}D} A_{PES}(\omega'). \tag{A.2}$$

The dash-dotted curves in Fig. A.2 show the result of this broadening for a standard deviation of 0.1 eV (left part of the figure) and 0.2 eV (right part of the figure), respectively. The standard deviation D was chosen to be in the same range as estimates of experimental resolution (which are around 0.1 for high resolution PES (Mo et al. 2003, Sekiyama et al. 2004) and approximately 0.2 to 0.36 for XAS (Inoue 2004, Müller et al. 1997)).

The comparison shows that the broadening has a strong effect especially on features with small width, i.e., the quasiparticle peak and the lower part of the upper Hubbard band. Extended features in the spectrum are practically not affected by the broadening. The quasiparticle peak in the XAS spectrum is reduced to half the height of the one in the PES spectrum due to the stronger broadening (0.2 Gaussian for XAS to 0.1 eV Gaussian for PES). Furthermore, the Fermi edge is distinctly sharper for the PES spectrum. One therefore has to take great care not to broaden the theoretical spectra too much, because important spectral features can be erroneously changed. Normally, we use the resolution of the experimental setup for the broadening of the theoretical spectrum. A good test to decide if the

[1] Usually, the width of a Gaussian with its infinite tails is given as the half width at half maximum (HWHM) which is $\sqrt{-2\ln\frac{1}{2}}D \approx 1.18D$ and corresponds to the experimental resolution.

broadening is correct is the comparison of the Fermi edge in theory and experiment. In Fig. A.3, this comparison is done for the experimental PES spectrum of $SrVO_3$ by Sekiyama et al. (2004) with our theoretical spectra multiplied with the Fermi function at the temperature of the experiments ($T = 20$ K) and broadened with different values $D = 0.05, 0.1$ and 0.2 eV. Obviously, the theoretical curve

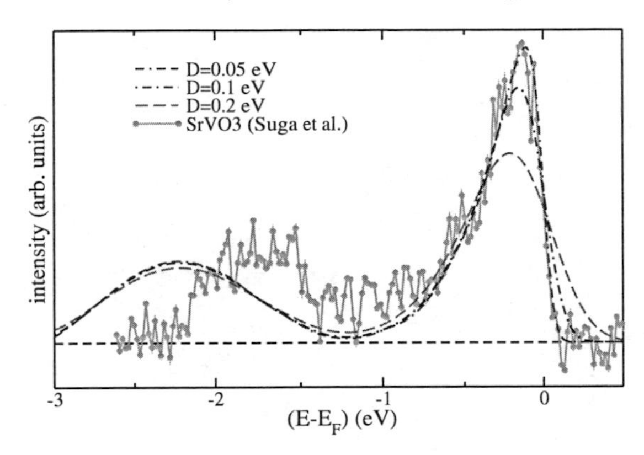

Figure A.3: PES Spectra for different broadenings D.

with $D = 0.05$ eV leads to the best agreement in the slope of the Fermi edge (and "accidentally" also in the height of the quasiparticle peak). The curve with the 0.1 eV broadening, i.e. with the resolution provided by the experimentalists, has a considerably shallower slope than the experimental curve. Therefore, the experimental estimate is probably too conservative and a smaller broadening should be used for the comparison. Where this was the case in our results of chapter 4, it is noted.

Finally, we want to clarify the effect of the temperature that is used in the Fermi function of equation A.1, especially in connection with broadening. Fig. A.4 shows spectra which were multiplied with the Fermi function at different temperatures. At higher temperatures ($\beta = 10 - 50$, i.e. $T \approx 1200 - 250$ K), the change of the quasiparticle height and that of the slope of the Fermi edge can be clearly seen, below $T \approx 250$ K, the difference is nearly negligible. When one further takes into account the Gaussian broadening (0.1 eV in figure A.5), the difference between curves at different temperatures is even smaller. Therefore it practically plays no role if one uses the Fermi function at the experimental temperature (which is between 20 K for the PES on $SrVO_3$ (Sekiyama et al. 2004) and 300 K for the XAS on V_2O_3 (Müller et al. 1997)) or at the temperature of the QMC calculations (which is 300 K for the data we compared with PES and XAS).

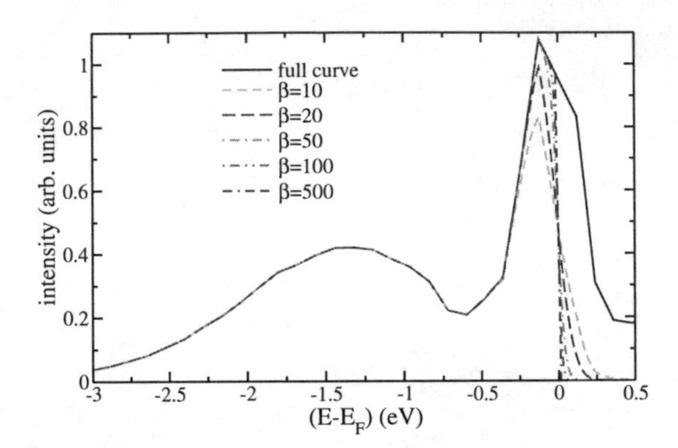

Figure A.4: Spectra multiplied with the Fermi function at different temperatures ($T \approx 1200 - 20$ K).

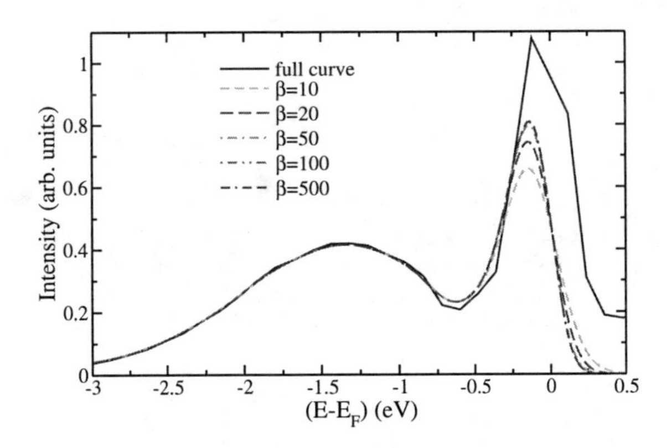

Figure A.5: Spectra multiplied with the Fermi function at different temperatures ($T \approx 1200 - 20$ K) and broadened with a Gaussian of width 0.1 eV.

BIBLIOGRAPHY

Aiura, Y., Iga, F., Nishihara, Y., Ohnuki, H. and Kato, H.: 1993, Effect of oxygen vacancies on electronic states of $CaVO_{3-\delta}$ and $SrVO_{3-\delta}$: A photoemission study, *Phys. Rev. B* **47**, 6732.

Allen, J. W.: 1976, Optical study of vanadium-ion interactions in $(V_2O_3)_x(Al_2O_3)_{1-x}$, *Phys. Rev. Lett.* **36**, 1249.

Allen, J. W.: 2004. private communication.

Andersen, O. K.: 1975, Linear methods in band theory, *Phys. Rev. B* **12**, 3060.

Andersen, O. K. and Jepsen, O.: 1984, Explicit, first-principles tight-binding theory, *Phys. Rev. Lett.* **53**, 2571.

Andersen, O. K., Saha-Dasgupta, T., Ezhov, S., Tsetseris, L., Jepsen, O., Tank, R. W., Arcangeli, C. and Krier, G.: 2001, Third-generation MTOs, *Psi-k Newsletter* **#45**, 86. http://psi-k.dl.ac.uk/newsletters/News_45/newsletter_45.pdf.

Anderson, P. W.: 1972, *Moment formation in solids*, Plenum Press, New York and London, p. 313. edited by W. J. L. Buyers.

Andres, K., Graebner, J. E. and Ott, H. R.: 1975, 4f-virtual-bound-state formation in $CeAl_3$ at low temperatures, *Phys. Rev. Lett.* **35**, 1779.

Anisimov, V. I., Aryasetiawan, F. and Lichtenstein, A. I.: 1997, First-principles calculations of the electronic structure and spectra of strongly correlated systems: the LDA+U method, *J. Phys. Cond. Matter* **9**, 767.

Anisimov, V. I., Kondakov, D. E., Kozhevnikov, A. V., Nekrasov, I. A., Pchelkina, Z. V., Allen, J. W., Mo, S.-K., Kim, H.-D., Metcalf, P., Suga, S., Sekiyama, A., Keller, G., Leonov, I., Ren, X. and Vollhardt, D.: 2005, Full orbital calculation scheme for materials with strongly correlated electrons, *Phys. Rev. B* **71**, 125119.

Anisimov, V. I., Korotin, M. A., Zölfl, M., Pruschke, T., Le Hur, K. and Rice, T. M.: 1999, Electronic structure of the heavy fermion metal LiV_2O_4, *Phys. Rev. Lett.* **83**, 364.

Anisimov, V. I., Poteryaev, A. I., Korotin, M. A., Anokhin, A. O. and Kotliar, G.: 1997, First-principles calculations of the electronic structure and spectra of strongly correlated systems: dynamical mean-field theory, *J. Phys. Cond. Matter* **9**, 7359.

Anisimov, V. I., Zaanen, J. and Andersen, O. K.: 1991, Band theory and Mott insulators: Hubbard U instead of Stoner I, *Phys. Rev. B* **44**, 943.

Arnold, D. J. and Mires, R. W.: 1968, Magnetic susceptibilities of metallic V_2O_3 single crystals, *J. Chem. Phys.* **48**, 2231.

Beach, K. S. D.: 2004, Identifying the maximum entropy method as a special limit of stochastic analytic continuation, *Preprint cond-mat/0403055* .

Berezin, F. A.: 1987, *Introduction to Superanalysis*, Reidel Dordrecht. edited by A. A. Kirillov.

Bethe, H.: 1931, Zur Theorie der Metalle. I. Eigenwerte und Eigenfunktionen der linearen Atomkette, *Z. Phys.* **71**, 205.

Bickers, N. E., Cox, D. L. and Wilkins, J. W.: 1987, Self-consistent large-N expansion for normal-state properties of dilute magnetic alloys, *Phys. Rev. B* **36**, 2036.

Bickers, N. E. and Scalapino, D. J.: 1989, Conserving approximations for strongly fluctuating electron systems. I. Formalism and calculational approach, *Ann. Phys.* **193**, 206.

Bickers, N. E. and White, S. R.: 1991, Conserving approximations for strongly fluctuating electron systems. II. Numerical results and parquet extension, *Phys. Rev. B* **43**, 8044.

Blankenbecler, R., Scalapino, D. J. and Sugar, R. L.: 1981, Monte Carlo calculations of coupled boson-fermion systems. I, *Phys. Rev. D* **24**, 2278.

Blümer, N.: 2002, *Mott-Hubbard Metal-Insulator Transition and Optical Conductivity in High Dimensions*, PhD thesis, Universität Augsburg.

Born, M. and Oppenheimer, J. R.: 1927, Zur Quantentheorie der Molekeln, *Ann. Physik* **84**, 457.

Brinkman, W. F. and Rice, T. M.: 1970, Application of gutzwiller's variational method to the metal-insulator transition, *Phys. Rev. B* **2**, 4302.

Brout, R.: 1960, Statistical mechanical theory of ferromagnetism: High density behavior, *Phys. Rev.* **118**, 1009.

Brown, P. J., Costa, M. M. R. and Ziebeck, K. R. A.: 1998, Paramagnetically aligned spin density in the metallic phase of V_2O_3: evidence for orbital exchange correlation, *J. Phys. Cond. Matter* **10**, 9581.

Brückner, W., Oppermann, H., Reichelt, W., Terukow, J. I., Tschudnowski, F. A. and Wolf, E.: 1983, *Vanadiumoxide: Darstellung, Eigenschaften, Anwendung*, Akademieverlag.

Bryan, R. K.: 1990, Maximum entropy analysis of oversampled data problems, *Eur. Biophys. J* **18**, 165.

Bulla, R.: 1999, Zero temperature metal-insulator transition in the infinite-dimensional Hubbard model, *Phys. Rev. Lett.* **83**, 136.

Bulla, R.: 2000, The numerical renormalization group method for correlated electrons, *Advances In Solid State Physics* **46**, 169.

Bulla, R., Hewson, A. C. and Pruschke, T.: 1998, Numerical renormalization group calculations for the self-energy of the impurity Anderson model, *J. Phys. Cond. Matter* **10**, 8365.

Burdin, S., Grempel, D. R. and Georges, A.: 2002, Heavy-fermion and spin-liquid behavior in a Kondo lattice with magnetic frustration, *Phys. Rev. B* **66**, 045111.

Byczuk, K. and Vollhardt, D.: 2002, Derivation of the Curie-Weiss law in dynamical mean-field theory, *Phys. Rev. B* **65**, 134433.

Caffarel, M. and Krauth, W.: 1994, Exact diagonalization approach to correlated fermions in infinite dimensions: Mott transition and superconductivity, *Phys. Rev. Lett.* **72**, 1545.

Capelle, K.: 2003, A bird's-eye view of density-functional theory, VIII'th Brazilian Summer School on Electronic Structure, p. 1. cond-mat/0211443.

Cardona, M. and Ley, L. (eds): 1978, *Photoemission in solids I: general principles*, Vol. 26 of *Topics in Applied Physics*, Springer, Berlin.

Carter, S. A., Rosenbaum, T. F., Metcalf, P., Honig, J. M. and Spalek, J.: 1993, Mass enhancement and magnetic order at the Mott-Hubbard transition, *Phys. Rev. B* **48**, 16841.

Castellani, C., Natoli, C. R. and Ranninger, J.: 1978, Magnetic structure of V_2O_3 in the insulating phase, *Phys. Rev. B* **18**, 4945.

Ceperley, D. M. and Alder, B. J.: 1980, Ground state of the electron gas by a stochastic method, *Phys. Rev. Lett.* **45**, 566.

Ceperley, D. M., Mascagni, M., Mitas, L., Saied, F. and Srinivasan, A.: 1998, The Scalable Parallel Random Number Generators library (SPRNG) for ASCI Monte Carlo computations, SPRNG, Version 1.0; http://sprng.cs.fsu.edu.

Chamberland, B. L. and Danielson, P. S.: 1971, Alkaline-earth vanadium (IV) oxides having the AVO_3 composition, *J. Solid State Chem.* **3**, 243.

Chmaissem, O., Jorgensen, J. D., Kondo, S. and Johnston, D. C.: 1997, Structure and thermal expansion of LiV_2O_4: Correlation between structure and heavy fermion behavior, *Phys. Rev. Lett.* **79**, 4866.

Cox, P. A.: 1992, *Transition Metal Oxides: An Introduction to their Electronic Structure and Properties*, Clarendon Press.

Dernier, P. D.: 1970, The crystal structure of V_2O_3 and $(V_{0.962}Cr_{0.038})_2O_3$ near the metal-insulator transition, *J. Phys. Chem. Solids* **31**, 2569.

Drchal, V., Janiš, V. and Kudrnovský, J.: 1999, *Electron Correlations and Material Properties*, Kluwer/Plenum, New York, p. 273.

Elfimov, I. S., Saha-Dasgupta, T. and Korotin, M. A.: 2003, Role of *c*-axis pairs in V_2O_3 from the band-structure point of view, *Phys. Rev. B* **68**, 113105.

Eyert, V.: 2000a. private communication.

Eyert, V.: 2000b, Basic notions and applications of the augmented spherical wave method, *Int. J. Quantum Chem.* **77**, 1007.

Eyert, V. and Höck, K.-H.: 1998, Electronic structure of V_2O_5: Role of octahedral deformations, *Phys. Rev. B* **57**, 12727.

Eyert, V., Höck, K.-H., Horn, S., Loidl, A. and Riseborough, P. S.: 1999, Electronic structure and magnetic interactions in LiV_2O_4, *Europhys. Lett.* **46**, 762.

Ezhov, S. Y., Anisimov, V. I., Khomskii, D. I. and Sawatzky, G. A.: 1999, Orbital occupation, local spin, and exchange interactions in V_2O_3, *Phys. Rev. Lett.* **83**, 4136.

Feldbacher, M., Held, K. and Assaad, F. F.: 2004, Projective quantum Monte Carlo method for the Anderson impurity model and its application to dynamical mean field theory, *Phys. Rev. Lett.* **93**, 136405.

Foex, M.: 1946, *Compt. Rend. Acad. Sci.* **223**, 1836.

Fujimori, A., Hase, I., Namatame, H., Fujishima, Y., Tokura, Y., Eisaki, H., Uchida, S., Takegahara, K. and de Groot, F. M. F.: 1992, Evolution of the spectral function in Mott-Hubbard systems with d^1 configuration, *Phys. Rev. Lett.* **69**, 1796.

Fujiwara, N., Yasuoka, H. and Ueda, Y.: 1998, Anomalous spin fluctuation in vanadium spinel LiV_2O_4 studied by ^{7}Li-NMR, *Phys. Rev. B* **57**, 3539.

Gebhard, F.: 1990, Gutzwiller correlated wave functions in finite dimensions d: A systematic expansion in $1/d$, *Phys. Rev. B* **41**, 9452.

Gebhard, F.: 1997, *The Mott metal-insulator transition*, Springer-Verlag, Berlin.

Georges, A. and Kotliar, G.: 1992, Hubbard model in infinite dimensions, *Phys. Rev. B* **45**, 6479.

Georges, A., Kotliar, G., Krauth, W. and Rozenberg, M. J.: 1996, Dynamical mean-field theory of strongly correlated fermion systems and the limit of infinite dimensions, *Rev. Mod. Phys.* **68**, 13.

Georges, A. and Krauth, W.: 1992, Numerical solution of the $d = \infty$ Hubbard model: Evidence for a Mott transition, *Phys. Rev. Lett.* **69**, 1240.

Goodenough, J. B.: 1970, Role of the crystal c/a ratio in Ti_2O_3 and V_2O_3, Proc. Tenth Intern. Conf. on the Physics of Semiconductors, U.S. Atomic Energy Commission, Oak Ridge, p. 304.

Goodenough, J. B.: 1971, *Metallic Oxides*, Vol. 5, Pergamon Press, Oxford, p. 145.

Gubernatis, J. E., Jarrell, M., Silver, R. N. and Sivia, D. S.: 1991, Quantum Monte Carlo simulations and maximum entropy: Dynamics from imaginary-time data, *Phys. Rev. B* **44**, 6011.

Gunnarsson, O., Andersen, O. K., Jepsen, O. and Zaanen, J.: 1989, Density-functional calculation of the parameters in the Anderson model: Application to Mn in CdTe, *Phys. Rev. B* **39**, 1708.

Gunnarsson, O., Jepsen, O. and Andersen, O. K.: 1983, Self-consistent impurity calculations in the atomic-spheres approximation, *Phys. Rev. B* **27**, 7144.

Gunnarsson, O. and Lundqvist, B. I.: 1976, Exchange and correlation in atoms, molecules, and solids by the spin-density-functional formalism, *Phys. Rev. B* **13**, 4274.

Gutzwiller, M. C.: 1963, Effect of correlation on the ferromagnetism of transition metals, *Phys. Rev. Lett.* **10**, 159.

Hammermesh, M.: 1989, *Group theory and its application to physical problems*, Dover Publ., New York. Reprint from Reading, Mass. 1962.

Han, J. E., Jarrell, M. and Cox, D. L.: 1998, Multiorbital Hubbard model in infinite dimensions: Quantum Monte Carlo calculation, *Phys. Rev. B* **58**, R4199.

Harrison, W. A.: 1980, *Electronic Structure and Properties of Solids*, Freeman, San Francisco.

Hedin, L. and Lundqvist, B. I.: 1971, Explicit local exchange-correlation potentials, *J. Phys. C: Solid State Phys.* **4**, 2064.

Held, K.: 1999, *Untersuchung korrelierter Elektronensysteme im Rahmen der Dynamischen Molekularfeldtheorie*, PhD thesis, Universität Augsburg. Shaker Verlag, Aachen, 1999.

Held, K.: 2000. private communication.

Held, K.: 2005. private communication.

Held, K., Allen, J. W., Anisimov, V. I., Eyert, V., Keller, G., Kim, H.-D., Mo, S.-K. and Vollhardt, D.: 2005, Two aspects of the Mott-Hubbard transition in Cr-doped V_2O_3, *Physica B* . Article in Press, cond-mat/0407787.

Held, K., Keller, G., Eyert, V., Anisimov, V. I. and Vollhardt, D.: 2001, Mott-Hubbard metal-insulator transition in paramagnetic V_2O_3: An LDA+DMFT(QMC) study, *Phys. Rev. Lett.* **86**, 5345.

Held, K., Nekrasov, I. A., Blümer, N., Anisimov, V. I. and Vollhardt, D.: 2001, Realistic modeling of strongly correlated electron systems, *Int. J. Mod. Phys. B* **15**, 2611.

Held, K., Nekrasov, I. A., Keller, G., Eyert, V., Blümer, N., McMahan, A. K., Scalettar, R. T., Pruschke, T., Anisimov, V. I. and Vollhardt, D.: 2003, Realistic investigations of correlated electron systems with LDA+DMFT, *Psi-k Newsletter*, Vol. #56, p. 65. http://psi-k.dl.ac.uk/newsletters/News_56/Highlight_56.pdf.

Held, K., Nekrasov, I., Keller, G., Eyert, V., Blümer, N., McMahan, A., Scalettar, R., Pruschke, T., Anisimov, V. and Vollhardt, D.: 2002, The LDA+DMFT approach to materials with strong electronic correlations, *in* J. Grotendorst, D. Marx and A. Muramatsu (eds), *NIC Series*, Vol. 10, Research Center Jülich, chapter Quantum Simulations of Complex Many-Body Systems: From Theory to Algorithms, p. 175. cond-mat/0112079.

Held, K. and Vollhardt, D.: 1998, Microscopic conditions favoring itinerant ferromagnetism: Hund's rule coupling and orbital degeneracy, *Eur. Phys. J. B* **5**, 473.

Hettler, M. H., Mukherjee, M., Jarrell, M. and Krishnamurthy, H. R.: 2000, Dynamical cluster approximation: Nonlocal dynamics of correlated electron systems, *Phys. Rev. B* **61**, 12739.

Hettler, M. H., Tahvildar-Zadeh, A. N. and Jarrell, M.: 1998, Nonlocal dynamical correlations of strongly interacting electron systems, *Phys. Rev. B* **58**, R7475.

Hewson, A. C.: 1993, *The Kondo Problem to Heavy Fermions*, Cambridge University Press, Cambridge.

Hirsch, J. E.: 1983, Discrete Hubbard-Stratonovich transformation for fermion lattice models, *Phys. Rev. B* **28**, 4059.

Hirsch, J. E. and Fye, R. M.: 1986, Monte Carlo method for magnetic impurities in metals, *Phys. Rev. Lett.* **56**, 2521.

Hohenberg, P. and Kohn, W.: 1964, Inhomogeneous electron gas, *Phys. Rev.* **136**, B864.

Hopkinson, J. and Coleman, P.: 2002a, LiV_2O_4: evidence for two-stage screening, *Physica B* **312**, 711.

Hopkinson, J. and Coleman, P.: 2002b, LiV_2O_4: Frustration induced heavy fermion metal, *Phys. Rev. Lett.* **89**, 267201.

Hubbard, J.: 1963, Electron correlations in narrow electron bands, *Proc. Roy. Soc. London A* **276**, 238.

Hubbard, J.: 1964a, Electron correlations in narrow electron bands II. The degenerate band case, *Proc. Roy. Soc. London A* **277**, 237.

Hubbard, J.: 1964b, Electron correlations in narrow electron bands III. An improved solution, *Proc. Roy. Soc. London A* **281**, 401.

Inoue, I. H.: 2004, private communication.

Inoue, I. H., Bergemann, C., Hase, I. and Julian, S. R.: 2002, Fermi surface of $3d^1$ perovskite $CaVO_3$ near the Mott transition, *Phys. Rev. Lett.* **88**, 236403.

Inoue, I. H., Goto, O., Makino, H., Hussey, N. E. and Ishikawa, M.: 1998, Bandwidth control in a perovskite-type $3d^1$-correlated metal $Ca_{1-x}Sr_xVO_3$. I. Evolution of the electronic properties and effective mass, *Phys. Rev. B* **58**, 4372.

Inoue, I. H., Hase, I., Aiura, Y., Fujimori, A., Haruyama, Y., Maruyama, T. and Nishihara, Y.: 1995, Systematic development of the spectral function in the $3d^1$ Mott-Hubbard system $Ca_{1-x}Sr_xVO_3$, *Phys. Rev. Lett.* **74**, 2539.

Inoue, I. H., Hase, I., Aiura, Y., Fujimori, A., Morikawa, K., Mizokawa, T., Haruyama, Y., Maruyama, T. and Nishihara, Y.: 1994, Systematic change of spectral function observed by controlling electron correlation in $Ca_{1-x}Sr_xVO_3$ with fixed $3d^1$ configuration, *Physica C* **235-240**, 1007.

Janiš, V. and Vollhardt, D.: 1992, Comprehensive mean field theory for the Hubbard model, *Int. J. Mod. Phys. B* **6**, 731.

Jarrell, M.: 1992, Hubbard model in infinite dimensions: A quantum Monte Carlo study, *Phys. Rev. Lett.* **69**, 168.

Jarrell, M.: 1997, Maximun entropy analytic continuation of quantum monte carlo data, *in* D. Scalapino (ed.), *Numerical Methods for Lattice Quantum Many-Body Problems*, Addison Wesley, Reading.

Jarrell, M., Akhlaghpour, H. and Pruschke, T.: 1993, Quantum Monte-Carlo in the infinite dimensional limit, *in* M. Suzuki (ed.), *Quantum Monte Carlo Methods in Condensed Matter Physics*, World Scientific, Singapore.

Jarrell, M. and Gubernatis, J. E.: 1996, Bayesian inference and the analytic continuation of imaginary-time quantum Monte Carlo data, *Physics Reports* **269**, 133.

Jarrell, M., Maier, T., Hettler, M. H. and Tahvildarzadeh, A. N.: 2001, Phase diagram of the Hubbard model: Beyond the dynamical mean field, *Europhys. Lett.* **56**, 563.

Jayaraman, A., McWhan, D. B., Remeika, J. P. and Dernier, P. D.: 1970, Critical behavior of the Mott transition in Cr-doped V_2O_3, *Phys. Rev. B* **2**, 3751.

Jepsen, O. and Andersen, O. K.: 2000, The STUTTGART TB-LMTO-ASA program, Max-Planck-Institut für Festkörperforschung, Stuttgart, Germany.

Jepson, O. and Anderson, O. K.: 1971, The electronic structure of h.c.p. Ytterbium, *Solid State Commun.* **9**, 1763.

Johnston, D. C.: 1999, Heavy fermion behaviors in LiV_2O_4, *Physica B* **281**, 21.

Johnston, D. C., Ami, T., Borsa, F., Crawford, M. K., Fernandez-Baca, J. A., Kim, K. H., Harlow, R. L., Mahajan, A. V., Miller, L. L., Subramanian, M. A., Torgeson, D. R. and Wang, Z. R.: 1995, *Ser. Solid State Sci.* **119**, 20.

Jones, R. O. and Gunnarsson, O.: 1989, The density functional formalism, its applications and prospects, *Rev. Mod. Phys.* **61**, 689.

Joo, J. and Oudovenko, V.: 2001, Quantum Monte Carlo calculation of the finite temperature Mott-Hubbard transition, *Phys. Rev. B* **64**, 193102.

Jung, M. H. and Nakotte, H.: n.d. unpublished.

Kanamori, J.: 1963, Electron correlation and ferromagnetism of transition metals, *Prog. Theor. Phys.* **30**, 275.

Keiter, H. and Kimball, J. C.: 1970, Perturbation technique for the Anderson Hamiltonian, *Phys. Rev. Lett.* **25**, 672.

Keller, G., Held, K., Eyert, V., Anisimov, V. I. and Vollhardt, D.: 2004, Electronic structure of paramagnetic V_2O_3: Strongly correlated metallic and Mott insulating phase, *Phys. Rev. B* **70**, 205116.

Knecht, C.: 2002, *Numerische Analyse des Hubbard-Modells im Rahmen der Dynamischen Molekularfeld-Theorie*, Diploma thesis, Universität Mainz.

Kohn, W. and Sham, L. J.: 1965, Self-consistent equations including exchange and correlation effects, *Phys. Rev.* **140**, A1133.

Kondo, S., Johnston, D. C. and Miller, L. L.: 1999, Synthesis, characterization, and magnetic susceptibility of the heavy-fermion transition-metal oxide LiV_2O_4, *Phys. Rev. B* **59**, 2609.

Kondo, S., Johnston, D. C., Swenson, C. A., Borsa, F., Mahajan, A. V., Miller, L. L., Gu, T., Goldman, A. I., Maple, M. B., Gajewski, D. A., Freeman, E. J., Dilley, N. R., Dickey, R. P., Merrin, J., Kojima, K., Luke, G. M., Uemura, Y. J., Chmaissen, O. and Jorgensen, J. D.: 1997, LiV_2O_4: A heavy fermion transition metal oxide, *Phys. Rev. Lett.* **78**, 3729.

Kotliar, G.: 1999, Landau theory of the Mott transition in the fully frustrated Hubbard model in infinite dimensions, *Eur. Phys. J. B* **11**, 27.

Kotliar, G., Lange, E. and Rozenberg, M. J.: 2000, Landau theory of the finite temperature Mott transition, *Phys. Rev. Lett.* **84**, 5180.

Kotliar, G., Savrasov, S. Y., Pálsson, G. and Biroli, G.: 2001, Cellular dynamical mean field approach to strongly correlated systems, *Phys. Rev. Lett.* **87**, 186401.

Kozhevnikov, A.: n.d. unpublished.

Krimmel, A., Loidl, A., Klemm, M., Horn, S. and Schober, H.: 1999, Dramatic change of the magnetic response in LiV_2O_4: Possible heavy fermion to itinerant d-metal transition, *Phys. Rev. Lett.* **82**, 2919.

Krimmel, A., Loidl, A., Klemm, M., Horn, S. and Schober, H.: 2000a, Interplay between spin glass and heavy fermion behavior in the d-metal oxides $Li_{1-x}Zn_xV_2O_4$, *Phys. Rev. B* **61**, 12578.

Krimmel, A., Loidl, A., Klemm, M., Horn, S. and Schober, H.: 2000b, Magnetic properties of the d-metal heavy-fermion system $Li_{1-x}Zn_xV_2O_4$, *Physica B* **276-278**, 766.

Kusunose, H., Yotsuhashi, S. and Miyake, K.: 2000, Formation of a heavy quasiparticle state in the two-band Hubbard model, *Phys. Rev. B* **62**, 4403.

Laad, M. S., Craco, L. and Müller-Hartmann, E.: 2003, *Phys. Rev. Lett.* **91**, 156402.

Lægsgaard, J. and Svane, A.: 1999, Theory of the alpha - gamma phase transition in Ce, *Phys. Rev. B* **59**, 3450.

Leung, T. C., Wang, X. W. and Harmon, B. N.: 1988, Band-theoretical study of magnetism in Sc_2CuO_4, *Phys. Rev. B* **37**, 384.

Levy, M.: 1979, Universal variational functionals of electron densities, first-order density matrices, and natural spin-orbitals and solution of the ν-representability problem, *Proc. Natl. Acad. Sci. (USA)* **76**, 6062.

Lichtenstein, A. I. and Katsnelson, M. I.: 1998, Ab initio calculations of quasiparticle band structure in correlated systems: LDA++ approach, *Phys. Rev. B* **57**, 6884.

Lichtenstein, A. I. and Katsnelson, M. I.: 2000, Antiferromagnetism and d-wave superconductivity in cuprates: A cluster dynamical mean-field theory, *Phys. Rev. B* **62**, 9283.

Lichtenstein, A. I., Katsnelson, M. I. and Kotliar, G.: 2003, Spectral density functional approach to electronic correlations and magnetism in crystals, *in* A. Gonis, N. Kioussis and M. Ciftan (eds), *Electron Correlations and Materials Properties 2*, Kluwer, New York, chapter Phenomological Studies of Correlation Effects. cond-mat/0211076.

Lieb, E. H. and Wu, F. Y.: 1968, Absence of Mott transition in an exact solution of the short-range, one-band model in one dimension, *Phys. Rev. Lett.* **20**, 1445.

Liebsch, A.: 2003a, Quasi-particle spectra of perovskites: Enhanced Coulomb correlations at surfaces, *Eur. Phys. J. B* **32**, 477.

Liebsch, A.: 2003b, Surface versus bulk Coulomb correlations in photoemission spectra of $SrVO_3$ and $CaVO_3$, *Phys. Rev. Lett.* **90**, 096401.

Liebsch, A.: 2004, Single Mott transition in multi-orbital Hubbard model. cond-mat/0405410.

Liechtenstein, A. I., Anisimov, V. I. and Zaanen, J.: 1995, Density-functional theory and strong interactions: Orbital ordering in Mott-Hubbard insulators, *Phys. Rev. B* **52**, R5467.

Limelette, P., Georges, A., Jérome, D., Wzietek, P., Metcalf, P. and Honig, J. M.: 2003, Universality and critical behavior at the Mott transition, *Science* **302**, 89.

Mahajan, A. V., Sala, R., Lee, E., Borsa, F., Kondo, S. and Johnston, D. C.: 2000, ^{7}Li and ^{51}V NMR study of the heavy-fermion compound LiV_2O_4, *Phys. Rev. B* **57**, 8890.

Maier, T., Jarrell, M., Pruschke, T. and Keller, J.: 2000, *d*-wave superconductivity in the Hubbard model, *Phys. Rev. Lett.* **85**, 1524.

Maiti, K., Sarma, D. D., Rozenberg, M. J., Inoue, I. H., Makino, H., Goto, O., Pedio, M. and Cimino, R.: 2001, Electronic structure of $Ca_{1-x}Sr_xVO_3$: A tale of two energy scales, *Europhys. Lett.* **55**, 246.

Makino, H., Inoue, I. H., Rozenberg, M. J., Hase, I., Aiura, Y. and Onari, S.: 1998, Bandwidth control in a perovskite-type 3d^1-correlated metal $Ca_{1-x}Sr_xVO_3$. II. optical spectroscopy, *Phys. Rev. B* **58**, 4384.

Matsuno, J., Fujimori, A. and Mattheiss, L. F.: 1999, Electronic structure of spinel-type LiV_2O_4, *Phys. Rev. B* **60**, 1607.

Mattheiss, L. F.: 1994, Band properties of metallic corundum-phase V_2O_3, *J. Phys. Cond. Matter* **6**, 6477.

McMahan, A. K., Held, K. and Scalettar, R. T.: 2003, Thermodynamic and spectral properties of compressed Ce calculated using a combined local-density approximation and dynamical mean-field theory, *Phys. Rev. B* **67**, 75108.

McMahan, A. K., Martin, R. M. and Satpathy, S.: 1988, Calculated effective Hamiltonian for La_2CuO_4 and solution in the impurity Anderson approximation, *Phys. Rev. B* **38**, 6650.

McWhan, D. B., Menth, A., Remeika, J. P., Brinkman, W. F. and Rice, T. M.: 1973, Metal-insulator transitions in pure and doped V_2O_3, *Phys. Rev. B* **7**, 1920.

McWhan, D. B. and Remeika, J. P.: 1970, Metal-insulator transition in $(V_{1-x}Cr_x)_2O_3$, *Phys. Rev. B* **2**, 3734.

Metropolis, N., Rosenbluth, A. W., Rosenbluth, M. N., Teller, A. H. and Teller, E.: 1953, Equation-of-state calculations by fast computing machines, *J. Chem. Phys.* **21**, 1087.

Metzner, W. and Vollhardt, D.: 1989, Correlated lattice fermions in $d = \infty$ dimensions, *Phys. Rev. Lett.* **62**, 324.

Mila, F., Shiina, R., Zhang, F.-C., Joshi, A., Ma, M., Anisimov, V. and Rice, T. M.: 2000, Orbitally degenerate spin-1 model for insulating V_2O_3, *Phys. Rev. Lett.* **85**, 1714.

Mizokawa, T. and Fujimori, A.: 1996, Electronic structure and orbital ordering in perovskite-type 3d transition-metal oxides studied by Hartree-Fock band-structure calculations, *Phys. Rev. B* **54**, 5368.

Mo, S.-K., Denlinger, J. D., Kim, H.-D., Park, J.-H., Allen, J. W., Sekiyama, A., Yamasaki, A., Kadono, K., Suga, S., Saitoh, Y., Muro, T., Metcalf, P., Keller, G., Held, K., Eyert, V., Anisimov, V. I. and Vollhardt, D.: 2003, Prominent quasiparticle peak in the photoemission spectrum of the metallic phase of V_2O_3, *Phys. Rev. Lett.* **90**, 186403.

Mo, S.-K., Kim, H.-D., Allen, J. W., Gweon, G.-H., Denlinger, J. D., Park, J.-H., Sekiyama, A., Yamasaki, A., Suga, S., Metcalf, P. and Held, K.: 2004, Filling of the Mott-Hubbard gap in the high temperature photoemission spectrum of $(V_{0.972}Cr_{0.028})_2O_3$, *Phys. Rev. Lett.* **93**, 076404.

Moeller, G., Si, Q., Kotliar, G., Rozenberg, M. J. and Fisher, D. S.: 1995, Critical behavior near the Mott transition in the Hubbard model, *Phys. Rev. Lett.* **74**, 2082.

Morikawa, K., Mizokawa, T., Kobayashi, K., Fujimori, A., Eisaki, H., Uchida, S., Iga, F. and Nishihara, Y.: 1995, Spectral weight transfer and mass renormalization in Mott-Hubbard systems $SrVO_3$ and $CaVO_3$: Influence of long-range Coulomb interaction, *Phys. Rev. B* **52**, 13711.

Morin, F. J.: 1959, Oxides which show a metal-to-insulator transition at the Neel temperature, *Phys. Rev. Lett.* **3**, 34.

Motome, Y. and Imada, M.: 1997, A quantum Monte Carlo method and its applications to multi-orbital Hubbard models, *J. Phys. Soc. Jap.* **66**, 1872.

Mott, N. F.: 1968, Metal insulator transition, *Rev. Mod. Phys.* **40**, 677.

Mott, N. F.: 1990, *Metal-Insulator Transitions*, 2nd edn, Taylor and Francis, London.

Müller-Hartmann, E.: 1989, Correlated fermions on a lattice in high dimensions, *Z. Phys. B* **74**, 507.

Müller, O., Urbach, J. P., Goering, E., Weber, T., Barth, R., Schuler, H., Klemm, M., Horn, S. and denBoer, M. L.: 1997, Spectroscopy of metallic and insulating V_2O_3, *Phys. Rev. B* **56**, 15056.

Nagaoka, Y.: 1966, Ferromagnetism in a narrow, almost half-filled s band, *Phys. Rev.* **147**, 392.

Negele, J. W. and Orland, H.: 1987, *Quantum Many-Particle Systems*, Addison-Wesley, New York.

Nekrasov, I. A., Held, K., Blümer, N., Poteryaev, A. I., Anisimov, V. I. and Vollhardt, D.: 2000, Calculation of photoemission spectra of the doped Mott insulator $La_{1-x}Sr_xTiO_3$ using LDA+DMFT(QMC), *Eur. Phys. J. B* **18**, 55.

Nekrasov, I. A. and Pchelkina, Z. V.: 2002. private communication.

Park, J.-H., Tjeng, L. H., Tanaka, A., Allen, J. W., Chen, C. T., Metcalf, P., Honig, J. M., de Groot, F. M. F. and Sawatzky, G. A.: 2000, Spin and orbital occupation and phase transitions in V_2O_3, *Phys. Rev. B* **61**, 11506.

Pavarini, E., Biermann, S., Poteryaev, A., Lichtenstein, A. I., Georges, A. and Andersen, O. K.: 2004, Mott transition and suppression of orbital fluctuations in orthorhombic $3d^1$ perovskites, *Phys. Rev. Lett.* **92**, 176403.

Pfalzer, P., Will, J., Nateprov Jr., A., Klemm, M., Eyert, V., Horn, S., Frenkel, A. I., Calvin, S. and denBoer, M. L.: 2002, Local symmetry breaking in paramagnetic insulating $(Al,V)_2O_3$, *Phys. Rev. B* **66**, 085119.

Pickett, W. E.: 1989, Electronic structure of the high-temperature oxide superconductors, *Rev. Mod. Phys.* **61**, 433.

Poteryaev, A., Lichtenstein, A. I. and Kotliar, G.: 2003, Non-local Coulomb interactions and metal-insulator transition in Ti_2O_3: a cluster LDA+DMFT approach. cond-mat/0311319.

Potthoff, M., Wegner, T. and Nolting, W.: 1997, Interpolating self-energy of the infinite-dimensional Hubbard model: Modifying the iterative perturbation theory, *Phys. Rev. B* **55**, 16132.

Di Matteo, S., Perkins, N. B. and Natoli, C. R.: 2002, Spin-1 effective Hamiltonian with three degenerate orbitals: An application to the case of V_2O_3, *Phys. Rev. B* **65**, 054413.

Pruschke, T.: 2004. private communication.

Pruschke, T., Cox, D. L. and Jarrell, M.: 1993, Hubbard model in infinite dimensions: Thermodynamic and transport properties, *Phys. Rev. B* **47**, 3553.

Pruschke, T. and Grewe, N.: 1989, The Anderson model with finite Coulomb repulsion, *Z. Phys. B* **74**, 439.

Reuter, B. and Jaskowsky, J.: 1960, LiV_2O_4-MgV_2O_4, a new spinel system with semiconducting properties, *Angew. Chem.* **72**, 209.

Rey, M. J., Dehaudt, P., Joubert, J. C., Lambert-Andron, B., Cyrot, M. and Cyrot-Lackmann, F.: 1990, Preparation and structure of the compounds $SrVO_3$ and Sr_2VO_4, *J. Solid State Chem.* **86**, 101.

Rice, T. M. and McWhan, D. B.: 1970, Metal-insulator transition in transition-metal oxides, *IBM J. Res. Develop.* **14**, 251.

Rozenberg, M. J.: 1997, Integer-filling metal-insulator transitions in the degenerate Hubbard model, *Phys. Rev. B* **55**, 4855.

Rozenberg, M. J., Chitra, R. and Kotliar, G.: 1999, Finite temperature Mott transition in the Hubbard model in infinite dimensions, *Phys. Rev. Lett.* **83**, 3498.

Rozenberg, M. J., Inoue, I. H., Makino, H., Iga, F. and Nishihara, Y.: 1996, Low frequency spectroscopy of the correlated metallic system $Ca_xSr_{1-x}VO_3$, *Phys. Rev. Lett.* **76**, 4781.

Rozenberg, M. J., Kotliar, G., Kajüter, H., Thomas, G. A., Rapkine, D. H., Honig, J. M. and Metcalf, P.: 1995, Optical conductivity in Mott-Hubbard systems, *Phys. Rev. Lett.* **75**, 105.

Rozenberg, M. J., Zhang, X. Y. and Kotliar, G.: 1992, Mott-Hubbard transition in infinite dimensions, *Phys. Rev. Lett.* **69**, 1236.

Rubtsov, A. N., Savkin, V. V. and Lichtenstein, A. I.: 2004, Continuous time quantum Monte Carlo method for fermions. cond-mat/0411344.

Sakai, S., Arita, R. and Aoki, H.: 2004, Numerical algorithm for the double-orbital Hubbard model: Hund-coupled pairing symmetry in the doped case, *Phys. Rev. B* **70**, 172504.

Sandvik, A. W.: 1998, Stochastic method for analytic continuation of quantum Monte Carlo data, *Phys. Rev. B* **57**, 10287.

Sandvik, A. W. and Scalapino, D. J.: 1995, Spin dynamics of La_2CuO_4 and the two-dimensional Heisenberg model, *Phys. Rev. B* **51**, 9403.

Savrasov, S. Y. and Kotliar, G.: 2001, Spectral density functionals for electronic structure calculations. cond-mat/0106308.

Schiller, A. and Ingersent, K.: 1995, Systematic $1/d$ corrections to the infinite-dimensional limit of correlated lattice electron models, *Phys. Rev. Lett.* **75**, 113.

Schramme, M.: 2000, *Untersuchungen zur Spektralfunktion von VO_2 und V_2O_3 mittels linear dichroischer Photoemission*, PhD thesis, Universität Augsburg.

Sekiyama, A., Fujiwara, H., Imada, S., Suga, S., Eisaki, H., Uchida, S. I., Takegahara, K., Harima, H., Saitoh, Y., Nekrasov, I. A., Keller, G., Kondakov, D. E., Kozhevnikov, A. V., Pruschke, T., Held, K., Vollhardt, D. and Anisimov, V. I.: 2004, Mutual experimental and theoretical validation of bulk photoemission spectra of $Sr_{1-x}Ca_xVO_3$, *Phys. Rev. Lett.* **93**, 156402.

Sham, L. J. and Kohn, W.: 1966, One-particle properties of an inhomogeneous interacting electron gas, *Phys. Rev.* **145**, 561.

Silver, R. N., Sivia, D. S. and Gubernatis, J. E.: 1990, Maximum-entropy method for analytic continuation of quantum Monte Carlo data, *Phys. Rev. B* **41**, 2380.

Singh, D. J., Blaha, P., Schwarz, K. and Mazin, I. I.: 1999, Electronic structure and heavy-fermion behavior in LiV_2O_4, *Phys. Rev. B* **60**, 16359.

Singhal, S. P.: 1975, Dielectric matrix for aluminium, *Phys. Rev. B* **12**, 564.

Skilling, J. and Bryan, R. K.: 1984, Maximum entropy image reconstruction: general algorithm, *Mon. Not. R. Astr. Soc.* **211**, 111.

Solovyev, I., Hamada, N. and Terakura, K.: 1996, t_{2g} versus all 3d localization in $LaMO_3$ perovskites (M=Ti-Cu): First-principles study, *Phys. Rev. B* **53**, 7158.

Stewart, G. R.: 1984, Heavy-fermion systems, *Rev. Mod. Phys.* **56**, 755.

Suzuki, M.: 1976, Relationship between d-dimensional quantal spin systems and (d+1)-dimensional Ising systems. equivalence, critical exponents and systematic approximants of the partition function and spin correlations, *Prog. Theor. Phys.* **56**, 1454.

Tanaka, A.: 2002, Electronic structure and phase transition in V_2O_3: Importance of 3d spin-orbit interaction and lattice distortion, *J. Phys. Soc. Jap.* **71**, 1091.

Trotter, H. F.: 1959, On the product of semigroups of operators, *Proc. Am. Math. Soc.* **10**, 545.

Ulmke, M.: 1995, *Phasenübergänge in stark korrelierten Elektronensystemen*, PhD thesis, KFA Jülich.

Ulmke, M., Janiš, V. and Vollhardt, D.: 1995, Anderson-Hubbard model in infinite dimensions, *Phys. Rev. B* **51**, 10411.

Urano, C., Nohara, M., Kondo, S., Sakai, F., Takagi, H., Shiraki, T. and Okubo, T.: 2000, LiV_2O_4 spinel as a heavy-mass fermi liquid: Anomalous transport and role of geometrical frustration, *Phys. Rev. Lett.* **85**, 1052.

van Dongen, P. G. J. and Janiš, V.: 1994, Mott transition near the ferromagnetic state, *Phys. Rev. Lett.* **72**, 3258.

Varma, C. M.: 1999, Heavy fermions in the transition-metal compound LiV_2O_4, *Phys. Rev. B* **60**, R6973.

Vlaming, R. and Vollhardt, D.: 1992, Controlled mean-field theory for disordered electronic systems: Single-particle properties, *Phys. Rev. B* **45**, 4637.

Vollhardt, D., Held, K., Keller, G., Bulla, R., Pruschke, T., Nekrasov, I. A. and Anisimov, V. I.: 2005, Dynamical mean-field theory and its applications to real materials, *J. Phys. Soc. Jap.* **74**, 136.

Vollhardt, D. and Wölfle, P.: 1990, *The Superfluid Phases of Helium 3*, Taylor & Francis.

von Barth, U. and Hedin, L.: 1972, A local exchange-correlation potential for the spin polarized case: I, *J. Phys. C: Solid State Phys.* **5**, 1629.

Wada, H., Shiga, M. and Nakamura, Y.: 1989, Low temperature specific heat of nearly ferro- and antiferromagnetic compounds, *Physica B* **161**, 197.

Wahle, J., Blümer, N., Schlipf, J., Held, K. and Vollhardt, D.: 1998, Microscopic conditions favoring itinerant ferromagnetism, *Phys. Rev. B* **58**, 12749.

Weiss, P.: 1907, L'hypothése du champ moléculaire et la propriété ferromagnétique, *J. de Phys.* **6**, 661.

Williams, A. R., Kübler, J. and Gelatt Jr., C. D.: 1979, Cohesive properties of metallic compounds: Augmented-spherical-wave calculations, *Phys. Rev. B* **19**, 6094.

Wilson, K. G.: 1975, The renormalization group: Critical phenomena and the Kondo problem, *Rev. Mod. Phys.* **47**, 773.

Wolenski, T.: 1999, *Combining bandstructure and dynamical mean-field theory: A new perspective on* V_2O_3, PhD thesis, Universität Hamburg. Shaker Verlag, Aachen, 1999.

Zaanen, J., Jepsen, O., Gunnarsson, O., Paxton, A. T., Andersen, O. K. and Svane, A.: 1988, What can be learned about high T_c from local density theory?, *Physica C* **153**, 1636.

Zaanen, J. and Sawatzky, G. A.: 1990, Systematics in band gaps and optical spectra of 3d transition metal compounds, *J. Solid State Chem.* **88**, 8.

Zaránd, G., Cox, D. L. and Schiller, A.: 2000, Toward a systematic 1/d expansion: Two-particle properties, *Phys. Rev. B* **62**, 16227.

Zölfl, M., Pruschke, T., Keller, J., Poteryaev, A. I., Nekrasov, I. A. and Anisimov, V. I.: 2000, Combining density-functional and dynamical-mean-field theory for $La_{1-x}Sr_xTiO_3$, *Phys. Rev. B* **61**, 12810.

LIST OF PUBLICATIONS

Parts of this thesis have already been published in several articles:

Held, K., Keller, G., Eyert, V., Anisimov, V. I. and Vollhardt, D.: "Mott-Hubbard metal-insulator transition in paramagnetic V_2O_3: An LDA+DMFT(QMC) study", *Phys. Rev. Lett.* **86**, 5345 (2001).

Held, K., Nekrasov, I., Keller, G., Eyert, V., Blümer, N., McMahan, A., Scalettar, R., Pruschke, T., Anisimov, V. and Vollhardt, D.: "The LDA+DMFT approach to materials with strong electronic correlations", *in* J. Grotendorst, D. Marx and A. Muramatsu (eds), *NIC Series*, Vol. 10, Research Center Jülich, chapter Quantum Simulations of Complex Many-Body Systems: From Theory to Algorithms, p. 175 (2002). cond-mat/0112079.

Mo, S.-K., Denlinger, J. D., Kim, H.-D., Park, J.-H., Allen, J. W., Sekiyama, A., Yamasaki, A., Kadono, K., Suga, S., Saitoh, Y., Muro, T., Metcalf, P., Keller, G., Held, K., Eyert, V., Anisimov, V. I. and Vollhardt, D.: "Prominent quasiparticle peak in the photoemission spectrum of the metallic phase of V_2O_3", *Phys. Rev. Lett.* **90**, 186403 (2003).

Held, K., Nekrasov, I. A., Keller, G., Eyert, V., Blümer, N., McMahan, A. K., Scalettar, R. T., Pruschke, T., Anisimov, V. I. and Vollhardt, D.: "Realistic investigations of correlated electron systems with LDA+DMFT", *Psi-k Newsletter*, Vol. #56, p. 65 (2003). http://psi-k.dl.ac.uk/newsletters/News_56/Highlight_56.pdf.

Nekrasov, I. A., Pchelkina, Z. V., Keller, G., Pruschke, T., Held, K., Krimmel, A., Vollhardt, D. and Anisimov, V. I.: "Orbital state and magnetic properties of LiV_2O_4", *Phys. Rev. B* **67**, 085111 (2003).

Keller, G., Held, K., Eyert, V., Anisimov, V. I. and Vollhardt, D.: "Electronic structure of paramagnetic V_2O_3: Strongly correlated metallic and Mott insulating phase", *Phys. Rev. B* **70**, 205116 (2004).

Sekiyama, A., Fujiwara, H., Imada, S., Suga, S., Eisaki, H., Uchida, S. I., Takegahara, K., Harima, H., Saitoh, Y., Nekrasov, I. A., Keller, G., Kondakov, D. E., Kozhevnikov, A. V., Pruschke, T., Held, K., Vollhardt, D. and Anisimov, V. I.:

"Mutual experimental and theoretical validation of bulk photoemission spectra of $Sr_{1-x}Ca_xVO_3$", *Phys. Rev. Lett.* **93**, 156402 (2004).

Anisimov, V. I., Kondakov, D. E., Kozhevnikov, A. V., Nekrasov, I. A., Pchelkina, Z. V., Allen, J. W., Mo, S.-K., Kim, H.-D., Metcalf, P., Suga, S., Sekiyama, A., Keller, G., Leonov, I., Ren, X., and Vollhardt, D.: "Full orbital calculation scheme for materials with strongly correlated electrons", *Phys. Rev. B* **71**, 125119 (2005).

Held, K., Allen, J. W., Anisimov, V. I., Eyert, V., Keller, G., Kim, H.-D., Mo, S.-K. and Vollhardt, D.: "Two aspects of the Mott-Hubbard transition in Cr-doped V_2O_3", *Physica B* **359**, 642 (2005).

Nekrasov, I. A., Keller, G., Kondakov, D. E., Kozhevnikov, A. V., Pruschke, T., Held, K., Vollhardt, D. and Anisimov, V. I.: "Comparative study of correlation effects in $CaVO_3$ and $SrVO_3$" (2005). cond-mat/0501240, accepted for Phys. Rev. B.

Vollhardt, D., Held, K., Keller, G., Bulla, R., Pruschke, T., Nekrasov, I. A. and Anisimov, V. I.: "Dynamical mean-field theory and its applications to real materials", *J. Phys. Soc. Jap.* **74**, 136 (2005).

CURRICULUM VITAE

Name	Georg Philipp Keller
Date of birth	26 April 1974
Place of birth	Füssen
Nationality	German

Education

09/80 – 07/84	Primary School: Grundschule Pfronten
09/84 – 06/93	Grammar School: Gymnasium Füssen
06/93	Abitur

Studies

11/93 – 11/99	Study of Physics and Computer Sciences at the Universität Augsburg
01/98 – 05/99	Diploma thesis under supervision of Prof. Dr. B. Rauschenbach
11/99	Diploma in Physics: "Simulation bei der Plasma-Immersions-Ionenimplantation"
01/00 – 06/05	Doctoral thesis under supervision of Prof. Dr. D. Vollhardt, Universität Augsburg

Employments

07/93 – 08/93	Military service in Füssen
04/96, 10/96	Working student at the Max-Planck-Institut für Plasmaphysik in Garching
since 01/00	Research assistant at the Chair of Theoretical Physics III, Universität Augsburg
since 06/00	System administrator at the Chair of Theoretical Physics III, Universität Augsburg

ACKNOWLEDGEMENTS

I specially thank Prof. Dr. Dieter Vollhardt for the possibility to work in his group on an exciting new topic on the forefront of solid-state research. He was always eager for discussions and gave me new impulses and suggestions for my research. Due to his support, I had the opportunity to attend numerous conferences and workshops.

I am grateful to Dr. Nils Blümer, Dr. Karsten Held and Prof. Dr. Thomas Pruschke, who always had an open ear for my questions and problems and helped me greatly with their knowledge of physical and numerical topics even after they had left Augsburg.

I further want to thank Dr. Ralf Bulla, Dr. Marcus Kollar, Ivan Leonov, Dr. Peter Pfalzer and Dr. Martin Ulmke for interesting discussions and Prof. Dr. Vladimir Anisimov and his group for a good and fruitful collaboration.

Thanks are also due to PD Dr. Volker Eyert and Dr. Igor Nekrasov for providing the LDA data for my calculations and answering my questions on density functional approach.

I thank Dr. Martin Schramme, Prof. Dr. James Allen and his group and Prof. Dr. Shigemasa Suga and his group for providing their experimental data for comparison with my theoretical results, as well as to the Leibniz Rechenzentrum in München and the John von Neumann-Insitut für Computing in Jülich for providing the supercomputer resources for the extensive calculations of this thesis.

My sincere thanks are due to all the members of the groups of Prof. Dr. Dieter Vollhardt and Prof. Dr. Arno Kampf for the very friendly and productive atmosphere at the center for electronic correlations and magnetism in Augsburg, among them PD Dr. Stefan Kehrein, Xinguo Ren, Michael Sekania and Dr. Robert Zitzler.

I also want to thank Dr. Karsten Held, Dr. Marcus Kollar and Prof. Dr. Thomas Pruschke for reading the drafts of my thesis and providing valuable comments and suggestions. Furthermore, I am grateful to Dr. Nils Blümer, Dr. Karsten Held, Dr. Marcus Kollar and Dr. Robert Zitzler for their LaTeX style files.

Finally, I want to thank Ilonka, who always believed in me, and my parents, my good friends and Wally, who supported me through a difficult time.